CONSERVATION IN THE BUILT ENVIRONMENT

CONSERVATION IN THE BUILT ENVIRONMENT

ROBERT D. PICKARD

LONGMAN

Addison Wesley Longman Limited,
Edinburgh Gate, Burnt Mill, Harlow,
Essex CM20 2JE, England
and Associated Companies throughout the world.

First published 1996

British Library Cataloguing-in-Publication Data
A catalogue record for this book is available from the British Library

ISBN 0-582-22818-2

Produced by Longman Singapore Publishers (Pte) Ltd.
Printed in Singapore

CONTENTS

PREFACE

This book has been written to provide guidance on the philosophical, practical, financial, legal, planning and policy issues associated with conservation in the built environment. So often these issues are looked at in isolation; however it is intended that this book, by linking various issues of relevance, provides an approach to the subject otherwise unavailable amongst current literature.

It is hoped that this text will prove an invaluable reference source for both practitioners and students. The latter will find the book relevant to undergraduate courses in town and country planning, the various surveying disciplines, land and building management, architecture, engineering and other courses leading to related professional qualifications. Moreover, the text has particular relevance to students on the growing number of specialist conservation and property-based post-graduate courses.

A wide range of information sources has been researched during the writing of this book, and details of the majority of these sources are provided at the end of each chapter. These details, and those of other specialist literature mentioned within the text, have been included to allow the reader to more easily undertake further study.

I have attempted to state the law and policy as correct at the beginning of 1996. The manuscript was completed in mid-1995; however, since then I have made a number of amendments to bring the information up to date to 1st February 1996. These amendments are included, with relation to conservation areas, in Appendix 6, and more generally in Appendix 7.

Rob Pickard
March, 1996

ACKNOWLEDGEMENTS

The author and publishers are grateful to the following for permission to reproduce copyright material:

The National Monuments Record for our Figs 4.1 & 5.2 (RCMHE Crown copyright); Richard Dennis Publications for our Fig. 5.1; Avon County Library (Bath Central Library) for our Figs 7.6 & 7.7; The Bath Chronicle for our Fig. 7.8; The Bath Archaeological Trust for our Figs 8.1, 8.2 & 8.3; The Society for the Protection of Ancient Buildings for our Appendix 1 (text of the SPAB Manifesto); ICOMOS UK for our Appendices 2 & 5 (text of the ICOMOS Venice and Lausanne Charters).

The author would like to thank Chartwell Heritage plc for their cooperation during the preparation of this book.

Chapter 1

LISTED BUILDINGS

Introduction

The system of listing buildings of special architectural or historic interest derives from the Town and Country Planning Act 1944. Since this Act various protective measures and controls have been made in relation to buildings contained in the statutory list. These are now consolidated in the provisions of the Planning (Listed Buildings and Conservation Areas) Act 1990 (P(LBCA) Act 1990). In 1992 the responsibility for maintaining the list and dealing with listing issues was handed to the Department of National Heritage (DNH) from the Department of Environment (DoE) which retains its responsibility in relation to planning control in conservation matters. The separate functions of these two departments were indicated in DNH Circular 1/92 and DoE Circular 20/92 respectively.

The definition of a listed building

A building of special architectural or historic interest

A building may be listed because it has 'special architectural or historic interest'. However, the statutory list, which includes a description of every listed building contained within it, is not provided to identify the special features for which a building has been listed. It merely includes a description of each building, including certain features which may or may not have special interest. In fact the list description has no legal standing if there is a challenge regarding the merits of the special architectural or historic features of a listed building. It has been deemed unnecessary to set down guidelines as to why a building should be listed, although guidance notes have been issued to English Heritage historic buildings inspectors to assist them in the process of selecting buildings for listing. It follows that a building is listed because expert opinion has decided that it has special character.

Before any work is carried out to a listed building it needs to be determined whether 'listed building consent' (LBC) will be required. Opinion should be sought from a local authority conservation officer, the relevant planning officer where there is no specialist officer, or in certain circumstances English Heritage, as to whether the works proposed are likely to affect the special architectural or historic character. The views of such persons may be challenged, but anyone proposing to do so should obtain independent opinion, such as from an architectural historian who is recognised to be an authority on the particular type of building under consideration.

The definition

In s. 1(5) of the P(LBCA) Act 1990 a listed building is defined as:

... a building which is for the time being included in a list compiled or approved by the Secretary of State under this section; and for the purposes of this Act –

(a) any object or structure fixed to the building;
(b) any object or structure within the curtilage of the building which, although not fixed to the building, forms part of the land and has done so since before 1 July 1948,

shall be treated as part of the building ...

The problem of defining such terms as 'building', 'structure', 'object', 'fixed to', 'within the curtilage of', is frequently the source of dispute. Parliament has chosen not to clarify the meaning of these terms. Nevertheless, by examining a number of policy determinations and court decisions it may be possible to give a clearer idea of what can be listed and what may form part of a listed building.

The building

Various *obiter* opinions have been given on the subject of what is a 'building' but the judiciary have not been able to reach a conclusive definition. In *Stevens v. Gourley* (1859) 76 B (NS) 99, Byles J. (p. 112) expressed the problem thus:

... The imperfection of human language renders it not only difficult, but absolutely impossible to define the word 'building' with any accuracy ...

In practice, the DoE (and now the Department of National Heritage) has relied on the definition which is now contained in s. 336(1) of the Town and Country Planning Act 1990. This states that a building includes ...

... any structure or erection and any part of a building, but not plant or machinery comprised in a building.

This means two things. First, that a building can be as large as a football stadium or as small as a boundary stone. Secondly, the listing applies to any part of the building. While an interior or an exterior matter may be the reason for a building being listed (i.e. the feature which creates the 'special interest') it is the whole building (i.e. all parts thereof) which is listed. An appeal case from 1983 (APP/5250/E/82/147 [1983] JPL 751) usefully illustrates this point in that a decision to refuse LBC for the removal of internal partitions to a listed public house in order to allow food servery was upheld because it would have resulted in the loss of a smoke-room and panelling and a reduction in the architectural or historic value of the whole building.

Two tests are used by the DNH to provide guidance on the terms 'structure' and 'erection' since these are not defined in the Town and Country Planning Act 1990. The first concerns whether the matter in question is three dimensional and the second looks at whether the item is securely fixed to the ground by means of a foundation or by some other method.

These tests enable a wide variety of objects to be listed (Figs 1.1 and 1.2). For instance, docks, cranes, bridges, telephone kiosks, railings, gatepiers, statues, gravestones and urns. All these satisfy both tests. However, in case disputes, while most objects satisfy the first test, the second test is more difficult (Ross, 1990). This will be considered in due course under the heading of 'Objects'.

According to s. 1(3) of the P(LBCA) Act 1990 the Secretary of State may assess the building itself and in respect of architectural or historic interest:

- how its exterior contributes to any group of buidings
- the desirability of preserving any man-made item fixed to the building or falling within its curtilage.

These are reasonably straightforward. Yet the inference that a listed building must be *man-made* is not strictly correct. For example, a boulder known as the 'Butterstone' near Cotherstone in Co. Durham was listed for its historical interest which derives from its use as a collection point to pass money and goods between people at the time of 'The Plague' (Fig. 1.3).

Government advice provided in 1987 concerning policy matters associated with historic buildings also indicated a number of unusual items which may form part of a listed building or may be listed in their own right including gazebos, temples, follies, grottoes, obelisks, park bridges, statues, urns, vases, ice houses, terraces, ha-has, crinkle crankle walls and boundary walls and gates and gate piers (DoE Circular 8/87).

Further clarification is given to English Heritage historic building inspectors who are provided with a manual of instructions indicating types of buildings which can be listed. Some of the more unusual buildings included in the manual are whipping posts, chairmen's boxes, raised pavements and sign posts.

Fig. 1.1 The historic environment of the River Tyne at Wallsend where a number of cranes, once associated with the shipbuilding industry, have been listed.

Fig. 1.2 The grade I listed High Level Bridge spanning the River Tyne between Newcastle and Gateshead and used on two levels for both road and rail traffic. It was designed by Robert Stephenson (probably aided by John Dobson) and was erected in 1849. The list entry describes it as 'one of the finest pieces of architectural iron work in the world'.

Fig. 1.3 The Butterstone situated near the village of Cotherstone in County Durham and listed grade II for its historical significance.

In practice, it is rare that a dispute arises over whether something can be listed in its own right. But problems can arise with the subsidiary part of the definition of a listed building with regard to *objects* and *structures.*

Objects

It is often when ownership of a listed building changes hands that disputes arise over whether an owner is able to sell or remove what is alleged to be a chattel. The critical test in relation to objects is whether they are 'fixtures'. This was established in the case of *Corthorn Land and Timber Co. v. Minister of Housing and Local Government* (1966) 17 P & CR 210 in which fifteenth-century carved wood-panelling was found to be part of the listed building. Not all fixtures to a building will require consent for removal, such as an ordinary light switch. In other words, the object must form part of the building's special character (for example, as a part of its architectural design) (Fig. 1.4) rather than merely being an ornament which has been added later. This follows from the application of common law tests of annexation. Even if there is a sufficient 'degree' of annexation, the 'purpose' of annexation must also be considered. (See *Holland v. Hodgson* (1872) LR 7CP 328 and *Leigh v. Taylor* [1902] AC 157.)

In this area of the law uncertainties occasionally arise over an object's perceived importance. In a test case concerning Orchardleigh Park in Somerset a gilt overmantel and a steel fender were found not to be fixtures to the listed house (APP/Q3305/E/89805215 [1991] JPL 116; [1992] JPL 706). Mendip District Council prosecuted against the removal of these items on the basis that they were part of the architectural fabric and history of a country house. The Secretary of State allowed an appeal with respect to a retrospective application for LBC following a ruling by the Crown Court. From an architectural and historic point of view the items were found not to be part of the original design of the building and therefore there was insufficient purpose of annexation for them to be protected under the listed building legislation.

The same approach with regard to common law principles must be taken in the case of objects found within the curtilage of a listed building. But the question of whether items such as garden statues found in the grounds of a stately home can become listed in their own right, or be a protected fixture, is not always clear.

The DNH's procedure in this area is based on the case of *Berkley v. Poulett* (1976) EC Digest 754 in which Lord Scarman (p. 763) concluded that:

> *... an object, resting on the ground by its own weight alone, can be a fixture, if it be so heavy that there is no need to tie it into a foundation and it were put in place to improve the realty. Prima facie, however, an object resting on the ground by its own weight is not a fixture ...*

To determine if an object can be regarded as a fixture to a listed building, or

Fig. 1.4 An object fixed to a listed building. Chimney piece in Sienna Scagolia work and containing Tuscan pillars and columns with capitals and plinths in white with ogee mouldings and anthemion motifs. It is situated in the library room at the grade I listed Belsay Hall in Northumberland, an early nineteenth-century Greek revival neo-classical hall, and was installed as part of an overall architectural scheme of design.

be listed in its own right, it is necessary to consider whether it was intended to remain *in situ*. In practice this depends upon such factors as size, weight, presence of foundations or its setting. Thus, if it would take a substantial effort to remove the object or if it could only be removed by demolition, it is likely to have a sufficient degree of annexation.

A useful example to illustrate this can be made regarding a case considered by the Parliamentary Commissioner for Administration (C278/V/430/J) concerning two ornamental urns on pedestals in the garden of a listed building which the DoE successfully argued had been correctly listed. The DoE sought legal advice and on the basis of *Bovis Ltd v. Secretary of State* (unreported) it was decided that there was a degree of permanence, suggesting an intention that the urns should remain *in situ* and could only be removed by a process which would amount to pulling them down, i.e. there was a sufficient degree of annexation. Apart from this, the DoE's inspector advised that the urns formed an integral part of the architectural design of the entrance to the house, i.e. there was a sufficient purpose to the annexation (Anthony, 1987).

Further clarification of policy in this area was indicated in a case concerning the removal of the statue of the Three Graces at Woburn Abbey ([1991] JPL 401). It was first considered that the statue should be regarded as part of the relevant listed building since the listed temple in which the statue stood and the statue's plinth were designed to display the statue permanently. Nevertheless the Secretary of State felt it wrong to use listed building enforcement procedures in relation to items which are generally considered to be chattels. Regard was had to the fact that the Three Graces was expressly commissioned from a sculptor of prominence before steps were taken to add it to the temple and that it had always been regarded as a major work of art.

The Three Graces case was followed by another similar Ministerial decision. This is a further aspect of the case relating to Orchardleigh Park as a sundial was removed from the terrace of the listed house. The sundial was not fixed to the terrace (except by its own weight) and had been moved from time to time to a different spot on the terrace. In fact it had been brought to Orchardleigh Park by its owner from his previous house. Thus it did not have a sufficient degree of permanence in its annexation and its purpose of annexation could not be construed as being part of the overall architectural scheme of the listed house. A similar conclusion was drawn in relation to artwork commissioned for the Time-Life Building in Bond Street, London, in an appeal decided in January 1995 (T/APP/F/93/x5990/630150–1 and T/APP/F/94/x5990/632377–8 [1995] JPL 241) (Carter, 1995).

It is clear that the common law rules regarding the removal of objects are not always easy to apply. In each case it is a question of fact and degree as to whether the removal of the item requires LBC. This problem is mirrored in the case of structures in which there are further issues which require clarification.

Structures

The leading cases in relation to 'fixed' structures reveal the significance of the purpose of annexation to the matter of listing.

In *Attorney General, ex rel Sutcliffe, Rouse, and Hughes v. Calderdale B.C.* (1982) 46 P & CR 399 a terrace of cottages was found to be part of a listed mill building as they were physically connected to it by means of a brick and stone bridge-like structure. However, in *Debenhams plc v. Westminster City Council* [1987] 1 All ER 51, a rating case concerning two buildings (one of which was listed in its own right, connected by a footbridge, the 'other' building was found not to fall within the listing), Lord Keith considered that the word 'structure' should be limited to such structures which are 'ancillary' to the building itself, for instance a stable block of a mansion or the steading of a farmhouse. The 'other' building was a modern building and did not have a historical connection with the listed building, unlike the terrace and mill. The matter of listing could not turn on the business use of the current occupier.

The same approach has been followed in relation to structures which 'form part of the land'. For example, in the case of *Watson-Smyth v. Secretary of State for the Environment and Cherwell D.C.* [1992] JPL 451, which concerned the removal of stones (formerly part of a ha-ha) that in turn had formed part of the garden boundary associated with the listed North Aston Hall, the ha-ha was found to be a structure, being 'an edifice or framework put together by framework or building' (p. 452). Furthermore, as the ha-ha was constructed as part of the layout of the grounds of the hall it was found to be an 'ancillary' structure and was within the curtilage of the hall. Thus the degree of annexation was quite clearly signified by the nature of the construction of the ha-ha and there was sufficient purpose to the annexation in that it formed part of the architectural scheme of the hall, which included its grounds comprising a formal garden extending to the ha-ha.

The issue of whether a structure is 'ancillary', in either the case of a fixture to a listed building or otherwise forming 'part of the land', is frequently considered alongside the matter of the curtilage. In law the definition of the curtilage is often problematical. Yet in practice, when dealing with listed buildings, finding a true definition is not as important as determining whether an object or structure lies 'within the curtilage' (Walker, 1992).

The curtilage

The classic definition of the curtilage which is commonly cited in planning litigation is to be found in the words of Lord Mackintosh in *Sinclair Lockhart's Trustees v. Central Land Board* (1950) 1 P & CR 195 (p. 204):

... ground which is used for the comfortable enjoyment of a house or other building ... although it has not been marked off or enclosed in any way. It is enough that it serves the purposes of the house or building in some necessary or useful way ...

Here it can be seen that the question of 'purpose' recurs. The statement suggests that it is not necessary to actually define the curtilage, but in order to decide whether land (including fixtures thereto) fall 'within' the curtilage of a building it must have some form of intimate relationship with the building.

This approach may be followed in the judgment of Buckley L.J. in *Methuen-Cambell v. Walters* [1979] 1 QB 525 (CA) (p. 543) which was referred to in both the *Calderdale* case and the *Watson-Smyth* case within the context of the curtilage:

... For one corporeal hereditament to fall within the curtilage of another, the former must be so intimately associated with the latter ... that the former in truth forms part and parcel of the latter ... This may extend to ancillary buildings ... and so forth. How far it is appropriate to regard this identity as parts of one messuage or parcel of land as extending must depend on the characteristic and circumstances of the items ...

Looking at the approach taken in cases concerning listed building control, the 'setting' of the building (in terms of special architectural or historic interest) as considered under s. 1(3) of the P(LBCA) Act 1990 is the main criterion for examining whether there is such an intimate relationship. Thus, in the *Calderdale* case the terraced mill houses were part of the architectural and historic setting of the listed mill and in the *Watson-Smyth* case the ha-ha was part of the historic layout of grounds designed with the hall. But in the *Debenhams* case there was no such relationship.

Unfortunately, in practice, determining whether an item falls 'within the curtilage' is not so simple. In an attempt to assist this process, Stephenson L.J. defined the curtilage (p. 409) as:

... an area of land which includes any related objects or structures which naturally form, or formed, with the listed building, an integral whole. The boundaries of the area are to be determined by such factors as may be relevant to the circumstances of the particular use and by the manner in which the listed building, any related objects or structures and the land have been or are being used ...

This is a very vague definition which does not provide much assistance. Of more help, Stephenson L.J. provided three tests to determine whether a structure would fall within the curtilage of a listed building:

- the physical 'layout' of the listed building and the structure
- their ownership, past and present
- their use or function past and present.

Thus in the *Calderdale* case the terraced mill houses had a close physical relationship with the mill, they had been in common ownership, and their use and function was as workers' houses which served the mill. However, the *Debenhams* case modified these tests by adding that a structure must be ancillary and subordinate to a listed building, or, in other words, the relationship must be one of an accessory to a principal. Moreover, to be ancillary the relationship would have had to have been akin to that of a 'steading to a farmhouse'. It would not turn on the business or manner of use of adjoining properties of a particular occupier.

In situations where the physical relationship could be said to be akin to a 'steading to a farmhouse' (test 1), the requirement of function or use (test 3) may depend on the question of ownership (test 2) as function and ownership freqently interrelate. The matter of ownership becomes a significant issue when a structure is sold for the purposes of providing a new use for land or buildings. The question then arises whether the structure remains 'within the curtilage' of a listed building and whether such a change will require listed building consent.

Structures, ownership and the curtilage

Disputes may arise when a structure associated with a listed building is sold as it may be argued that a new curtilage is created thereby taking the structure out of listed building control. The courts have had some difficulty in dealing with this problem.

In the *Calderdale* case, the mill and houses had originally been in the same ownership and, as such, had a functional relationship. Thus the mill and houses were originally within the same curtilage. But at the time of listing there had been a fragmentation of ownership. All of the houses in the terrace had been sold, with the exception of the first house which was directly fixed to the bridge, in the year before the mill was listed. Nevertheless, despite the change in ownership, the houses were found to fall under the listing provisions because at some point in the past their function and ownership had been interdependent. Stephenson L.J. determined this on the basis that, in his view, the whole terrace formed *one structure* which was linked to the mill to make an integral whole. In the *Debenhams* case, Lord Keith, considering the concept of principal and accessory, confirmed that the whole terrace could be regarded as an ancillary structure. Unfortunately he did not deal with the aspect of ownership (Pickard, 1993).

Two further cases have considered this aspect. First, *Watts v. Secretary of State* [1991] JPL 718, in which part of an estate (including a manor house and barns and other farm buildings) were sold prior to the listing of the manor house and large barn (as detailed in the list description). The original owner of the estate retained a small barn (to convert to a separate dwelling) and an adjacent wall which were not mentioned in the list description. The owner carried out works to insert a door into the wall. It was argued that LBC was not necessary for this as the property was in wholly independent use.

The Deputy Judge considered that there would be little difficulty in finding a wall to be a structure ancillary to a listed building if at the time of listing it served the purpose of securing the building or its curtilage as an accessory to the principal building. He further stated that for a structure to be 'ancillary' (or an 'accessory') it must be subordinate to the principal listed building in both a physical and functional sense. But it was concluded that at the time of listing there was no functional connection as the wall then formed part of the curtilage of a building separate from the listed building in terms of ownership and physical occupation. Thus the wall did not serve the purposes of the listed building as it was not within its curtilage, nor was it a 'fixed' structure within the 'ancillary concept' as it was not subservient and subordinate to the listed building in *both* a physical and functional sense *at the time of listing.*

This case can be differentiated from the *Calderdale* decision as one house which formed part of the whole structure was in the same ownership as the mill at the time of listing. The Deputy Judge did, however, express concern that problems could arise if the question of ownership was given too much emphasis in deciding the issue. This follows the views of Stephenson L.J. in the *Calderdale* case in that he had stated that the strict conveyancing position was not necessarily of significance to what lies within the curtilage of a listed building.

In the second case, *R. v. London Borough of Camden, ex parte Bellamy* [1992] JPL 225, a decision to grant planning permission to extend a former coach house was challenged by way of judicial review. The former coach house had originally been in the same ownership as a house, and separated from it by a common garden. The house was listed in 1950 but the coach house had been sold in 1970 and had been used as an architect's office from 1966. The court did not decide on the listed building issue, being a case of review, but gave direction to the local planning authority (LPA). Nolan J. indicated that if the coach house had been used in connection with the house at the time of listing there would have been a sufficient physical and functional connection for it still to be regarded as an ancillary, independent, structure within the curtilage of the listed house. Thus despite the strict conveyancing position, the change in ownership would not bar the coach house from being included in the listing provided that at the time of listing there was a sufficient functional connection between it and the house.

The time of listing appears to have great significance. In *Watts* the structure in question, although fixed to the listed building, could not be regarded as ancillary as at the time of listing it functionally served another building. In the *Bellamy* case it was found that the converse would apply if it could be deduced that there was a functional connection at the time of listing. In both cases the question of ownership at the date of listing seems to have some relevance in providing a functional connection.

Using the time of listing as a marker for determining disputes in this area can be drawn from another field of listed building law. The case of *Robbins v.*

Secretary of State for the Environment and Another [1989] EGCS 35 concerned a Repairs Notice which identified works for restoring a building to its former glory. However, the House of Lords imposed a limitation on the repairs specified in the notice to such works as would restore the building to its condition at the time of listing and not to that at some period of time before this date.

Summary of considerations

The definition of the listed building itself is wide and includes any 'structure' or 'erection' and may even include a natural feature. But it is not generally with regard to the actual listed building that definition disputes arise. The problem area is where objects or structures are removed or altered without the necessary consent. This frequently happens where there is a change of ownership. The common law rules regarding the degree and purpose of annexation are significant when determining whether objects or structures are fixtures to a listed building or form part of the land within the curtilage. These are both determined by fact and degree. In this respect the degree of annexation is usually a simple issue dependent on matters such as size, weight, and the relative permanency of fixing. However, the purpose of annexation is more difficult to determine.

The question of purpose relates to whether some object or structure forms an integral part of the overall architectural or historic scheme of design with a listed building. In the case of a structure, it must be ancillary to a listed building. To be ancillary there must be a relationship of an accessory to a principal.

Sir Graham Eyre (as deputy Judge) observed *obiter* in the *Watts* case that if the structure, at the time of listing, secured the listed building or its curtilage in a physical and functional manner there would be a relationship of an accessory to a principal. The *Watts* case would appear to provide a sensible way through this rather difficult problem. Moreover, the *Bellamy* case and the decision of the House of Lords in the *Robbins* case add weight to the date of listing as being the determining time when the question of the physical and functional relationship should be examined. This approach has been followed in two subsequent appeal decision (T/APP/F/91/z3825/607134/p6 [1992] JPL 1084 and T/APP/L5810/E/91/808553/p5 [1993] JPL 602) concerning issues similar to those raised in the *Watts* case.

An alternative to resolve the problem of what forms part of a listed building is, as Lord Keith stated in the Debenhams case, to list 'that which is worthy of listing' (p. 59). It should be noted that Lord Montague (the former Chairman of English Heritage), in a speech to the House of Lords in 1986 (*Hansard* H.L. Deb., 13 October 1986, col. 623) referred to improvements in listing practices, made in light of the *Calderdale* case, which would leave little room for doubt as to what a listing includes. Moreover, government advice issued jointly by the DNH and DoE in 1994 through the publication of Planning Policy Guidance Note 15: Planning and the Historic Environment (PPG 15) indicates at para. 3.33 that the Secretary of State for National Heritage 'has attempted to consider individually all structures and buildings on a site which can be construed as separate buildings and to list those which qualify for listing' following the resurvey of buildings of architectural and historic interest since 1970. Yet, the evidence of court decisions indicate that the issue of what is included in a listing is still far from clear.

At the same time the courts have recognised the difficulties of defining what comes under listed building control; for instance, the Deputy Judge in the *Bellamy* case expressed sympathy for the parties and all planning authorities for this 'potentially recurring problem' (p. 263). Thus, it may be suggested that the list descriptions should be more explicit.

At first glance the idea of having a statutory list which defines exactly what items come under listed building control and the area of the land within the curtilage of a building has much to commend it (Evans, 1992). Yet it could lead to an all or nothing approach which could result in important features not being protected. For example, a hidden wall fresco discovered in the course of repair works to listed council houses in Westminster in 1991 led to the Museum of London taking a photographic record before protecting and reconcealing it prior to the building's rehabilitation (Construction News, 1992). Such a discovery is likely to add to the features which create the 'special interest' for which a building has been listed in the first place.

The fact that the list provides a description but does not identify the special characteristics for which a building has been listed allows some scope for the protection of unknown features, which may add to the 'special architectural or historic interest'. If there is doubt as to whether something falls within the scope of listing it is open to the proposer of change to seek advice from the LPA (Mynors, 1993). Paragraphs 3.35 of PPG 15 provide considerations which may assist LPAs in forming their own views or in giving advice. These include:

- the historical independence of the building;
- the physical layout of the principal building and other buildings;
- the ownership of the buildings now and at the time of listing;
- whether the structure forms part of the land;
- the use and function of the buildings, and whether a building is ancillary or subordinate to the principal building.

Although it must be remembered that any opinion offered by a planning or conservation officer may not be binding on the LPA (Rutherford *et al.*, 1986). The safest route is most probably to seek the views of the Secretary of State's advisers, namely English Heritage. Of note, English Heritage has argued for the defining of factors which make up 'special interest', but the use of legally binding management contracts in this context is not yet possible in law (see Appendix 7).

The listing procedure

The criteria for listing

The purpose of the system of protection under which buildings are listed is to enable control, according to certain criteria, over change to an identified stock of buildings. In considering whether to list a building, the Secretary of State may, under s. 1(3) of the P(LBCA) Act 1990, take into account not only the building itself but also two other factors:

- Any respect in which its exterior contributes to the architectural interest *or* historic interest of any group of buildings of which it forms part.

  For example, in *Iveagh v. Minister of Housing and Local Government* [1964] JPL 395, it was decided that it was proper to consider the interest of a whole terrace in deciding whether to list a building forming part of the terrace.
- The desirability of preserving, on the ground of its architectural interest *or* historic interest, any feature of the building or a man-made object or structure fixed to the building or forming part of the land comprised within the curtilage.

  For example, in the *Debenhams* case it was decided that the building connected to the actual listed building by means of a footbridge was historically independent and not part of the listing.

There is no judicial interpretation of what constitutes 'architectural' or 'historic' interest (Suddards, 1988) but *Guidance Notes for those Concerned in the Survey of Listed Buildings* published by the DoE in 1985 suggested that an assessment of 'historic interest' should deal with:

- the importance of the person or event, and
- the importance of the building in relation to that person's life and work or that event.

The factors considered to make up historic interest have since been refined by para. 6.10 of PPG 15 to include 'buildings which illustrate important aspects of the nation's social, economic, cultural or military history' and may also involve close historical associations 'with nationally important people or events'. Paragraph 6.15 of PPG 15 further indicates that well-documented historical associations of national importance may increase the case for listing and may justify a higher grading than would otherwise be appropriate. An example of a building listed almost entirely for its historical association is that of 'Underhill' in Gateshead (Fig. 1.5). The building is listed Grade II* but it is a very ordinary Victorian house. However, it was listed because Sir Joseph Swan, FRS, who invented the incandescent electric light bulb, was a tenant of the property between 1869 and 1883, the house being the first to be wired for domestic electric lighting.

Architectural interest is meant to have a wide meaning and will include types of materials, building expression and form, mass, scale, space, colour,

Fig. 1.5 Underhill, 99 Kells Lane, Gateshead. Sir Joseph Swan, the inventor of the incandescent light, lived here from 1869–83. The house was the first in England to be wired for domestic electric lighting and is listed grade II* for this historical association.

light, etc., associated with a building (Suddards, 1988). Paragraph 6.10 of PPG 15 identifies architectural interest in buildings worthy of listing as encompassing national importance in terms of 'architectural design, decoration and craftsmanship' and may also refer to 'important particular building types and techniques (e.g. buildings displaying technological innovation or virtuosity) and significant plan forms'. Furthermore, the external appearance of a building in terms of its intrinsic architectural quality and any group value are key considerations for assessing the aesthetic merits for listing purposes.

However, the trend has moved in the direction of a multi-disciplinary approach making it difficult to draw a line between architectural *and* historic interest as separate issues. A good example to understand this may be made in relation the industrial revolution settlement of Saltaire. The church, almshouses, public buildings and model housing in Italianate style have added significance, being part of the philanthropist Sir Titus Salt's pioneer settlement provided for the workers in his storeyed textile mill (Cherry, 1993).

Section 1(1) of the P(LBCA) Act 1990 sets out the basic test for a building's listability (i.e. does it possess special architectural or historic interest?). This is a discretionary matter which can be tested in the courts against certain *principles of selection* which are set out in paras 6.10–6.12 of PPG 15. The Secretary of State is also required by s. 1(4) of the Act to consult English Heritage and 'such other persons or bodies as appear to him appropriate as having special knowledge of, or interest in, buildings of architectural or historic interest' but is not obliged to heed what they say.

The *principles of selection* should be read in conjunction with more detailed standards which were adopted by the DoE in the accelerated resurvey of listed buildings carried out in the 1980s. These standards, which cover four time periods, were originally drawn up by the former Historic Buildings Council and are now specified in para. 6.11 of PPG 15:

- *All buildings built before 1700 which survive in anything like their original condition are listed.* In practice this standard relates to timber-framed buildings or buildings of other materials where external or internal features have survived. The relative completeness of the original building, age and scarcity are also important factors (Fig. 1.6).

- *Most buildings of about 1700 to 1840 are listed, though some selection is necessary.* Here greater significance is given to eighteenth-century buildings than to those originating from the beginning of the nineteenth century. In the case of Georgian town houses the pre-1770 period is more important although post-1770 buildings may be listed if they have external surviving features of architectural interest. Group value, prominence of siting and any historical associations can also be important considerations.

- *Between 1840 and 1914 greater selection is necessary to identify the best examples of particular types, and only buildings of definite quality and character are listed. and the selection is designed to include the works of the principal architects.* Quality

Fig. 1.6 Bessie Surtees House, Sandhill, Newcastle upon Tyne. Erected in the 16th and 17th century with eighteenth- and twentieth-century alterations, timber framed with Jacobean ceilings and chimney pieces, listed grade I; now occupied by English Heritage as their Northern Region office.

and character can be in one particular aspect, e.g. the architect, the design, decoration, quality of workmanship, planning, social or technological interest, etc. Buildings designed by the principal architects from this period have usually been selected for listing. These include such notable names as the gothic revivalist G.W. Pugin, P. Webb and W.G. Lethaby (founder members of the Society of the Protection of Ancient Buildings), R. Norman Shaw and Sir Edward Lutyens among others. Works of lesser-known architects from this period can also be considered for listing (Fig. 1.7).

Fig. 1.7 Grade II listed Lodge to the Banquetting Hall, Jesmond Dene, Newcastle upon Tyne. Designed by R. Norman Shaw, one of the principal architects during the period 1840–1914.

- *After 1914 only selected buildings of definite quality and character are normally listed.* From this period a body of examplars has been built up including Battersea Power Station and Stratford Memorial Theatre.

Following the recommendations set out in the House of Commons Environment Committee (First Report) on Historic Buildings in 1987, outstanding buildings originating after 1939 are now eligible for listing. This criterion for selection is based on a rolling programme, meaning that the building must have been constructed 30 years before it is considered for listing. The first such listing was Bracken House, the former Financial Times Building, which was constructed between 1955 and 1959. In fact, buildings between 10 and 30 years old may now be listed where they are of acknowledged international importance and they are under threat (Ross, 1991).

However, all but 18 of the 70 recommendations for listing put forward by English Heritage in 1987 under the new standard were rejected by the DoE. Following this, in 1992, English Heritage embarked on a research programme to identify key exemplars of post-1939 buildings which may be suitable for listing. The first 47 buildings (including 95 separate structures), comprising university and school buildings, recommended for listing from this investigation were accepted by the new department responsible for listing matters, the DNH, in March 1993.

Some of the new listings have included buildings of traditional styles such as Nuffield College in a Cotswold vernacular style and the neo-classical Queen's College and Lady Margaret Hall at Oxford. Others have been listed for their uncompromising modernism, including Leicester University's Engineering Building which has cantilevered lecture theatres and an outstanding glazed workshop roof. Examples of new building technology have been listed, including four *system-built* schools in Hertfordshire and the first school to be erected using the *CLASP* system at Mansfield. Many of the new listings are at grade II* – with the exceptions of St Catherine's College, Oxford, designed by the Danish architect Arne Jacobsen, and Falmer House, the first permanent building erected at the University of Sussex, designed by Sir Basil Spence, which were listed at grade I (Kay, 1993).

The second major property sector to be scrutinised for post-1939 listings was that of commercial, retail, industrial and railway architecture. By 1993 three examples had been listed including Time & Life building, London, designed by Michael Rosenauer in the early 1950s. This was listed grade II in 1988 for its architecture combined with integral works of art (Murdoch, 1993). In March 1995, 40 post-war buildings were identified as possible cases for listing, including the controversial Centre Point office block in London (1961–65), the Woolworths shop in Canterbury (1954), the Bank of England Printing Works in Debton (1954), Rugby Signal Box (1964) and Norgas House in Killingworth Township, Northumberland (1963–65) (Fig. 1.8). Plans to examine a wide range of building types – including public and private housing, churches, public buildings, buildings associated with health

Fig. 1.8 Norgas House, Killingworth Township constructed 1963–5. One of 40 modern buildings proposed for listing in 1995 but subject to public consultation under new arrangements.

care, transport, communications, the military and the navy, and planned towns – may result in further post-war buildings being listed but this may depend on the outcome of a public consultation exercise on the 1995 proposals (Worsley, 1995; Sudjic, 1995; Boseley, 1995).

From the above, it can be deduced that the principal criterion for listing remains the age of the building, and combined with this is its state of preservation when surveyed and the scarcity within a particular building type. Locality is also an important factor – the quality of buildings may vary throughout the country, but it is the DNH's policy to ensure that there is a proper spread of listed buildings throughout the country.

As well as the date groups considered above, the DNH's leaflet *What Listing Means: A Guide for Owners and Occupiers* published in 1990 suggests other considerations which may be relevant in choosing buildings for listing:

- *Special value within certain types.* This is either for architectural or planning reasons or to illustrate social and economic history (e.g. hospitals, railway stations, theatres, almshouses, mills, markets and industrial buildings).
- *Technological innovation or virtuosity.* Here, documentary evidence will usually be required for providing the case for listing on this basis. Examples include buildings of cast iron prefabrication or which represent the early use of reinforced concrete (Fig. 1.9).
- *Association with well-known characters or events.* Here, documentary evidence will usually be required. While association is a matter of degree, a strong association is usually necessary.
- *Group Value, especially as examples of town planning.* Examples given for this category include squares, terraces, and model villages (Fig. 1.10). Such buildings may be identified in the statutory list by the abbreviation 'GV'. As difficulties have arisen from time to time over listing on this ground, the DoE's 1985 leaflet entitled *Guidance Notes for those Concerned in the Survey of Listed Buildings* provided the following special advice:

  > *... The replacement factor must also be kept in mind when selecting buildings in the group category. A building must not be chosen merely because its neighbours are good and one is afraid that if it were demolished it would be replaced by an incongruous monstrosity. The design of new buildings can and should be controlled under general planning powers and it is not permissible therefore to place a building on the statutory list simply to ensure a congruity of neighbourhood which could as well be achieved in a new building. It must be possible to say that the existing building has some quality in relation to the context which no new building could have ...*

A further note was provided on reasons for selecting buildings from the inter-war period for listing. This indicated the range of buildings considered and the quality of the individual buildings actually selected. This was to give

Fig. 1.9 The C.W.S. Warehouse, Quayside, Newcastle upon Tyne; grade II listed. The construction (during 1891–1900) represents an early use of reinforced concrete, using the *Hennebique* method, with the exterior finished to resemble Portland stone. This building was identified as being 'at risk' in 1990 but has since become part of a regeneration scheme for the East Quayside.

recognition of the varied architectural styles in the period which were broadly to be interpreted as: modern, classical and others. Nine building categories were identified:

- churches, chapels and other places of worship;
- cinemas, theatres, hotels and other places of public entertainment;
- commercial and industrial premises, including shops and offices;
- schools, colleges and educational buildings;
- flats;

Fig. 1.10 Buildings of the model village of Ford, Northumberland built in the 1860s; listed *c.* 1914 for group value. Some of the gatepiers and garden walls are individually listed and are included in the statutory list for their contribution to group value.

- houses and housing types;
- municipal and other public buildings;
- railway stations, airport terminals and other places associated with public transport;
- miscellaneous.

The principles for selecting post-1939 buildings will probably be published following the completion of English Heritage's research programme and consideration by the Secretary of State of the further recommendations for listing according to building types. It is uncertain whether this will encompass new principles for all post-1914 buildings. However, it is most likely that the following matters will be relevant in determining the suitability of twentieth-century buildings for listing:

- *architectural composition* as reflected in the relative proportions of a building or the relationship of solid to void;
- *spacial treatment and scale* often revealed in modern designs by the creation of interpenetrating internal and external spaces via the use of large windows;
- *architectural idiom or style* such as in the earliest examples of new movements in design, for instance, *New Brutalism* or *Post-Modernism*;
- *intelligence, ingenuity or innovation* in the planning and siting of a buiding as well as in the use of structure and services;
- *materials* in terms of quality and ingenuity in choice and use;
- *integrity* with respect to the building as a total design via consistency in its various components, structure, services, planning, architectural form and detailing.

Other factors may also be relevant. For instance, historical associations and examples of rarity do not stop at a particular moment in time and therefore must be considered in more recent buildings (Judd, 1993; English Heritage, 1993).

The listing process

The basis for listing is set out in s. 1(1) of the P(LBCA) Act 1990. In practice, while the Secretary of State is under a duty to complete lists, most of the groundwork is undertaken by English Heritage historic buildings inspectors. Under s. 1(4) of the P(LBCA) Act 1990 the Secretary of State *must* consult English Heritage and 'such other persons or bodies of persons as appear to him appropriate as having specialist knowledge of, or interest in, buildings of architectural or historic interest'. The Secretary of State is not obliged to heed what they say and any decision is final. However, listing decisions can be tested in the courts by way of judicial review if it is felt that the Secretary of State has acted unreasonably.

An example of an action for judicial review may be found in the case of *Amalgamated Investment and Property Co. Ltd v. John Walker and Sons Ltd* [1976]

3 All ER 509 in which it was debated whether an owner of a building should be consulted regarding the fact that it had been proposed for listing. There is, however, no duty placed upon the Secretary of State to consult owners, occupiers or anyone else. The *John Walker* case did set out the mechanics of the listing process, which is also described in a number of other references (Suddards, 1988; Ross, 1991) and, since this decision, list survey workers have been bound by a code of conduct. Thus there is now a need for general publicity before an area-wide survey is conducted.

This practice was not extended to actually notifying an intention to list particular buildings. However, the procedure was changed in March 1995 when the Secretary of State for National Heritage announced that new proposals to list 40 post-war commercial, retail, industrial and railway buildings would be subject to public consultation (Boseley, 1995; Worsley, 1995). The practice which existed up to this date had been to follow the strict statutory duty contained under s. 2(3) of the P(LBCA) 1990 to notify owners and occupiers and the local authority when a listed building proposal has been approved. As a new listing takes effect on the date the list is signed, it has frequently taken some time before an owner or occupier is informed. Yet lack of knowledge of a listing has not saved a person from having to comply with the relevant statutory protections. This may be indicated by the case of *R. v. Wells Street Metropolitan Stipendary Magistrate, ex parte Westminster City Council* [1986] JPL 903 in which a builder was fined for carrying out works to a building on the instructions of the owner in ignorance of the fact that the building was listed. The new procedure will alert owners of an intention to list and should assist in resolving this potential problem in the future.

The main method for listing buildings is by way of a general resurvey of all buildings in an area. The last resurvey was commenced in 1970 and resulted in a four-fold increase in the number of listed buildings following 24 years' work although, as indicated earlier, English Heritage is now surveying the stock of post-1939 buildings according to particular types. Paragraph 6.6 of PPG 15 explains the relative gradings of listed buildings as follows:

Grade I: Buildings considered to be of exceptional interest (about 2 per cent, or 9000 listed buildings).
Grade II*: Important buildings of more than special interest (about 4 per cent, or 18,000 listed buildings).
Grade II: Buildings of special interest which warrant every effort to preserve them (about 94 per cent or 413,000 listed buildings).

Spot listing

Apart from listings arising out of a general resurvey or survey of a particular period, buildings may be listed by a procedure known as *spot listing*. This will normally be used in one of three circumstances (Ross, 1991):

- The building is in a unresurveyed area but conforms to current listing criteria and is under a perceived threat.
- The building is in a resurveyed area but was overlooked because it is hidden within an area of undistinguished buildings or some fresh evidence of its special interest has been discovered.
- The building is in a resurveyed area but fresh knowledge or developing tastes have altered the perception of its special interest.

The system of spot listing is invariably used to give consideration to the merit of listing buildings which may have special interest or are under threat from redevelopment proposals. It is open to any member of the public to bring to the DNH's attention a building thought to be under threat, and local authorities have been requested to be vigilant in redevelopment areas so that, if necessary, a special emergency listing can be arranged if demolition is thought to be imminent.

The spot-listing procedure is sometimes thought to be too slow to protect buildings of special interest. An example may be provided in the now infamous case of the Art Deco Firestone factory in Brentford which, as reported in *The Sunday Telegraph*, 1 August 1980, was demolished on a bank holiday before the mechanism of the spot-listing procedure could be brought into action.

However, there is also the danger that the desire to protect may lead to action which cannot be sustained if proper consideration is not given to the question of a building's purported special interest. A recent example may be indicated in the case of a telephone exchange in the City of London (Sands, 1991). The owners of the building, the Japanese Development Company Daiwa Securities, had paid £100 million to secure this site for the purposes of demolishing the telephone exchange and building their new European headquarters. The spot listing was a significant blow for the company whose proposed replacement had received admiration from the Royal Fine Art Commission and the City Corporation. The action prompted Diawa Securities to employ the architectural historian Professor Roy Worskett to write a report on the merits of the building. The outcome of this was that while the building was of interest, a better example of the same building type by the same architect was to be found nearby. It was concluded that the building should not remain listed (Ayers and Moreton, 1992).

Since the completion of the resurvey of listed buildings, the spot-listing procedure is arguably not going to have frequent use in the future. However, with the current research being carried out regarding post-1939 buildings it is perhaps with respect to this period that the procedure will be most frequently used. Moreover, the proposals to extend listing to modern office buildings and retail centres is likely to result in action to spot list buildings subject of refurbishment or redevelopment proposals in areas of high investment potential. By example, the former Sanderson headquarters in London was listed while planning negotiations were under way for major redevelopment proposals. Owners may be able to test the likelihood of listing in such situations by submitting a token planning application (Murdoch, 1993).

Deemed listing

Apart from the spot-listing procedure, local planning authorities (LPAs), other than a county planning authority, or, in London, English Heritage or the appropriate London Borough Council, can extend a temporary listing to all buildings under s. 3 of the P(LBCA) Act 1990 by means of a Building Preservation Notice. (Ecclesiastical buildings still in ecclesiastical use are excepted from this provision.) The principal reasons for using this approach are indicated as:

- if it appears to the LPA that a building is of special architectural or historic interest, or
- if a building is in danger of demolition or alteration in such a way as to affect its character in terms of its considered special interest.

The temporary listing, which has the effect of protecting the building as if it were listed for a period of six months, may or may not be confirmed by the Secretary of State at the end of the period of six months. This period is used for consideration as to the merits of the purported special interest in the building. If the listing is confirmed it is formally added to the statutory list. According to s. 3(3) of the P(LBCA) Act 1990 a Building Preservation Notice comes into force as soon as it has been served on both the owner and the occupier of the building to which it relates. However, under s. 4 of the Act, if there is a perceived urgency to protect a building the relevant notice may be affixed conspicuously to some object on the building. This provision was first enacted in 1972 to overturn a ruling in the case of *Maltglade Ltd v. St Albans R.D.C.* [1972] 1 WLR 1230.

Immunity from listing

Under s. 6(1) of the P(LBCA) Act 1990, where an application for planning permission has been made or granted for development involving the alteration, extension or demolition of a building, any person may apply to the Secretary of State for a certificate that he does not intend to list the building. This mechanism was provided following the *John Walker* case so that redevelopment proposals may not be thwarted or delayed at a late stage by the listing of a building subject of a development scheme. It allows an opportunity, if a certificate is granted, to fulfil development proposals without the threat of a possible listing for a period of five years. Further information on this facility was provided in a leaflet published by the DoE in 1987 entitled *A*

Note about Certificates of Immunity from Listing. This matter is now dealt with by the Listing Branch of the DNH. It should be noted that the relevant certificate also prevents a building from becoming subject of a Building Preservation Notice (as well as listing) for a period of five years.

According to para. 43 of DoE Circular 8/87 the process of applying for the relevant certificate would involve a complete reassessment of a building, and although the replacement advice contained in paras 6.28–6.33 of PPG 15 has not referred to this, it should be assumed that a reassessment of the building will be carried out. The law is at present silent on the position remaining if in fact the Secretary of State decides not to issue a certificate, but it can be fairly stated that such a decision is likely to be followed by action to add a building to the statutory list. This was the case in relation to Bracken House, the first post-1939 building to be listed under the 30-year rule.

Appeals against listing

There is no statutory basis for appealing against a building being placed on the statutory list. However, listed buildings can, nevertheless, be de-listed in certain circumstances:

1. First, where the listing was a mistake. For example, it was reported in *The Times*, 11 April 1988, that the DoE had listed what was thought to be a sixteenth-century house in Norfolk which was actually constructed between 1983 and 1988.
2. Secondly, where the building does not have any special interest. This may be evidenced by the case involving the proposal to spot list Daiwa's telephone exchange building mentioned earlier, which was overturned because it was not, as first thought, the only example by the particular architect and in fact there were better examples which could be listed if the Secretary of State so chose.
3. Lastly, the building may have been altered since it was listed. This may include buildings which have been demolished with or without consent or buildings which have been altered so as to lose their special interest.

However, the case of *R. v. Leominster District Council, ex parte Antique Country Buildings Ltd and Others* [1988] 2 PLR 23 (which involved the dismantling of a barn for export purposes) reveals that if sufficient components of the original building are still extant, it will fall to be a 'building' and the listing provisions will continue to apply. In each case it will be a matter of fact and degree. In the *Leominster* case 70–80 per cent of the structural members were extant and thus the LPA were entitled to issue a listed building enforcement notice requiring the restoration of the listed barn.

Furthermore, where a building has been altered – for instance, to provide a new use for an obsolete listed building – then although relevant policies advise for allowing new uses to safeguard long-term preservation, the very action of alteration may cause a building to lose the special interest for which it was listed in the first place. This was evidenced in the 1990 English Heritage statement entitled *The Conversion of Historic Farm Buildings* (updated in 1993) which highlighted examples of farm buildings which have been de-listed following their conversion to residential use.

In addition, while there is no statutory right of appeal against listing, it should be noted that s. 21(3) of the P(LBCA) Act 1990 provides reasons for an appeal against the refusal of LBC on the grounds that the building should not have been listed in the first place. However, para. 6.27 of PPG 15 indicates that the Secretary of State will not normally entertain an application for de-listing if the building is subject of an application for LBC or an appeal against refusal of such consent, or if action is being pursued concerning unauthorised works or neglect.

There are no set procedures for making an appeal against listing but the DoE produced a useful guidance note on the matter in 1988 entitled *How to Appeal Against Listing*. Requests to de-list buildings now have to be directed to the DNH.

Understanding the list

Copies of the statutory list of listed buildings are made available for public inspection at the Historic Buildings and Monuments Commission (English Heritage) in London and copies of relevant areas are held by LPA's. Public libraries also hold copies for inspection. But otherwise copies of the list are not generally available for purchase or distribution.

The various geographical sections of the list are commonly known as *greenbacks* and usually cover a town or local authority area and may be divided up according to relevant parishes. When an area is resurveyed, or when a spot listing is made, additional list entries may be added to the greenbacks to ensure that the list available for public inspection is up to date.

List entries

Each list entry has two sections: The statutory information and the description.

The statutory information

This is the top section of the list entry which includes information in three columns.

Left column	*Centre column*	*Right column*
OS sheet reference	Local Authority (name)	Street,town, village, or settlement name
DNH reference		Building name, number, or other identification
Date of listing		List grade

The left column may also include the initials 'GV' if there is any group value. Moreover, buildings may be grouped for a list entry such as for a whole terrace, square or other group of buildings. List entries for a number of properties within the same street may follow according to odd numbers and even numbers. Where a corner building is listed it may have two addresses which will most likely be cross-referenced. Alternative addresses or names of buildings may also have to be checked on the statutory map which accompanies the 'greenback'.

The description

Below the statutory information lies a description of the property which normally follows the mnemonic *B DAMP FISHES*. This can be explained as follows:

B – Building type (original use followed by any new use)
D – Date of construction and date of alterations
A – Architect (if there is no named architect the relevant craftsman or patron may be named)
M – Materials used in the structure, cladding, decoration and roof
P – Plan or style of layout
F – Façade description
I – Interior features
S – Subsidiary features such as boundary walls, gatepiers or garden features
H – Historical information such as associations with well-known characters or events
E – Extra information such as in relation to a building's setting or its connection with a famous collection of art or furniture
S – Sources of information from which the description has been compiled.

The consequences of listing is that the relevant LPA or the Secretary of State 'shall have regard to the desirability of preserving the building or its setting or any features of special architectural or historic interest which it possesses'. This is a broad statement, the effects of which will be considered in the following chapters.

References

Anthony, B. (1987) What is a building?, *Conservation Bulletin*, Issue 3, October, pp. 6–7.

Ayres, J. and Moreton, D. (1992) Spot listing: disease or cure?, *The Estates Gazette*, February, pp. 98 and 110.

Boseley, S. (1995) Railtrack signals row over listings, *The Guardian*, 15 March, p. 10.

Carter, H. (1995) Planning and the historic environment, *Conservation Bulletin*, Issue 25, March, pp. 18–19.

Cherry, M. (1993) Listing criteria: 'historic interest', *Conservation Bulletin*, Issue 19, March, pp. 19–20.

Construction News (1992) Project report, renovation reveals fine detailing, *Construction News Products Supplement*, January/February, p. 28.

DoE Circular 8/87 (1987) Historic Buildings and Conservation Areas – Policy and Procedures.

English Heritage (1993) General Principles for the Selection of Post-1939 Educational Buildings for Listing.

Evans, W. (1992) What counts as listed [1992] JPL 432.

Hansard (1986) House of Lords, 13 October, col. 626.

Judd, M. (1993) History catches up with present, *The Estates Times*, 28 May, p. 10.

Kay, D. (1993) Post-war listing: An update, *Conservation Bulletin*, Issue 20, July, pp. 12–13.

Murdoch, A. (1993) Optimum solutions for listed buildings, *The Estates Gazette*, 29 May, pp. 90 -1.

Mynors, C. (1993) The extent of listing [1993] JPL 99.

Pickard, R.D. (1993) Listed buildings – the continuing dilemmas of definition, *Conveyancer and Property Lawyer*, May/June, pp. 192–205.

Ross, M. (1991) *Planning and the Heritage: Policy and Procedures*, E. & F.N. Spon.

Rutherford, L., Peart, J.D. and Pickard, R.D. (1986) Estoppel and development control counter service [1986] JPL 891.

Sands, D. (1991) DOE cuts off Daiwa phone listing, *The Estates Gazette*, 29 June, p. 67.

Suddards, R.W. (1988) *Listed Buildings*, 2nd edn, Sweet & Maxwell.

Sudjic, D. (1995) Blessing a blot on the landscape, *The Guardian*, 15 March, p. 22.

Walker, R. (1992) Objects and structures associated with listed buildings. Paper presented to the Historic Houses Association conference on Listed Building Controls, 15 April, London.

Worsley, G. (1995) Listed: the right to make money, *The Independent on Sunday*, 12 March, p. 12.

Chapter 2

PLANNING AND LEGAL CONTROLS OVER LISTED BUILDINGS

Introduction

Under the system of town and country planning a listed building has a special protection in that works to the building may require LBC and unauthorised work may result in a listed building enforcement notice (LBEN). These matters are additional to the normal planning requirements which may affect a listed building under development control policy and law. Until the issuance of PPG 15 in 1994, there was a presumption in favour of the preservation of a listed building. Although the preservation of listed buidings remains a fundamental aspect of government policy, there are now more general criteria against which applications for LBC should be considered. Furthermore, the setting of a listed building and its contribution to its local scene is a material consideration in deciding whether to grant planning permission for new development which may have an impact upon this factor. The procedures associated with the submission of applications associated with works to, or development proposed in the setting of, a listed building, or appeals against refusals of consent, requires particular scrutiny in this context.

Apart from these aspects of planning control there are other powers available to the relevant authorities which may be used to secure the preservation of a listed building. These include procedures to secure the condition of a listed building where it is in a state of disrepair (see Chapter 4) and to reinstate the condition following unauthorised works, backed up by criminal penalties.

New responsibilities for conservation matters

The creation of the Department of National Heritage in 1992 resulted in a division of responsibilities in the administration of conservation matters between the new department and the Department of the Environment. These were indicated in Circular 20/92 (DoE) and Circular 1/92 (DNH).

With respect to listed buildings the DNH deals with the procedures associated with the listing of buildings as well as repairs notices, urgent works notices and other associated land acquisition matters.

The DoE has responsibility for planning control matters including the call-in of LBC applications, decisions on called-in applications, on applications by local planning authorities, and on appeals against refusals of LBC and development by other government departments. It also deals with all related enforcement, modification/revocation, purchase notice and compensation procedures.

Listed building control

The requirement to obtain listed building consent (LBC) for works associated with listed buildings was first provided under the Town and Country Planning Act 1968. This provision is now found under s. 7 of the P(LBCA) Act 1990 which states that no person is permitted to carry out 'works for the demolition of a listed building or for its alteration or extension in any manner which would affect its character' unless LBC has been first obtained.

Control over demolition and alterations

The control of both these aspects is of great concern to property owners, occupiers and advisers particularly as, in legal terms, it is not entirely clear what constitutes 'demolition' or 'works ... which would affect its character'. Moreover, it can be difficult to determine a dividing line between these two issues. The inference from the above statutory provision is that LBC will be required for (a) the demolition of a listed building, and (b) alterations and extensions which may affect the character of a listed building. The drafting of this provision is somewhat ambiguous but most legal commentators seem to accept that in the case of 'alterations and extensions' consent will only be required where they affect the 'character' of a listed building (Pickard, 1993b). Thus in a Crown Court case concerning purported unauthorised works to a listed house in Beverly ([1991] JPL 705) the jury found that the removal of eighteenth-century walling from the ground floor had constituted an offence whereas an additional charge concerning the removal of a ground-floor staircase was not pursued because, in itself, it was not considered to be of architectural or historic merit.

Whether the removal of an item such as a staircase will require LBC will inevitably depend on the item being a fixture, which may in turn depend on the item forming part of the scheme of design for the building (see Chapter 1). However, a more difficult issue to contemplate is that an alteration will almost certainly include some demolition. This has significance as applications for LBC which concern demolition are arguably more strictly controlled with a requirement that the Royal Commission on the Historical

Monuments of England (RCHME) are afforded access to record a building prior to its demolition, and six national amenity societies, which have been designated as statutory consultees, are required to be notified of the proposals. These are:

- the Ancient Monuments Society
- the Council of British Archaeology
- the Georgian Group
- the Society for the Protection of Ancient Buildings
- the Victorian Society
- the Twentieth Century Society (formerly known as the Thirties Society, receives notifications via the Victorian Society)

Furthermore, the Secretary of State must be informed of cetain demolition applications. PPG 15, compared to its policy predecessor, is less specific on the type of demolition works that may be called-in by the Secretary of State. However, until new directions are provided for the handling of LBC applications the advice contained in para. 86 of DoE Circular 8/87 concerning demolition is still relevant. Thus the Secretary of State must be notified of all applications for LBC for the demolition of grade I and grade II* buildings and in relation to *works of demolition* to grade II buildings. There are eight categories of application that may involve *works of demolition* to grade II buildings:

1. The total or substantial demolition of a principal building.
2. The total demolition of a curtilage building (unless it is expressly stated in the list description that it is not of special interest).
3. The demolition of part of a principal building where the cubic content, taken together with any other demolished part since the building was listed, exceeds 10 per cent of the cubic content of the building when listed (subject to certain date exceptions).
4. The total demolition of an elevation of a principal building.
5. The demolition of substantially all of the interior of a principal building.
6. The partial or total demolition of any object or structure fixed to a principal building or a curtilage building (unless it is expressly stated in the list description that it is not of special interest).
7. The total or partial demolition of a building where the application is made within five years of the Secretary of State's determination of a previous such application.
8. Applications affecting buildings which have received or which are the subject of a grant application under the Historic Buildings and Ancient Monuments Act 1953.

Despite the specifiction of these categories of application it still remains unclear, in legal terms, as to what actual works amount to demolition or alteration. The courts have considered this problem but have not resolved it. For instance, in *R. v. North Hertfordshire D.C., ex parte Sullivan* [1981] JPL 752,

Comyn J. (p. 753) stated that 'not every piece of work by way of alteration or extension necessarily amounted to demolition'. Unfortunately the court did not attempt to clarify the distinctions in terminology. Furthermore, in *Shimizu (UK) Ltd v. Westminster City Council* [1995] 23 EG 118, the Court of Appeal determined by majority that the Lands Tribunal had erred in law in deciding that the removal of internal chimney breasts in a listed building were alterations. However, the dissenting judgment of Russell L.J. was that the works did not amount to demolition but to an alteration to part of the building in both the vertical and horizontal planes.

To assist in this matter, Annex C of PPG 15 provides some more detailed guidance on alterations which will require LBC for the purpose of assisting LPAs in deciding on the suitability of proposed changes to listed buildings. However, this still does not clarify what constitutes works of demolition.

Academic debate on the vexed question of what is demolition has not reached any conclusive position (Mynors, 1992; Bold, 1992). Yet the removal or replacement of windows, doors, or small sections of interior panelling or plasterwork are likely to constitute alterations, whereas the removal or replacement of a whole scheme of interior finishes or a boundary wall, or a substantial part of a building, are likely to constitute demolition. In other words, each case much be decided on its own merits. Without doubt the national amenity societies and other bodies with conservation interests monitor applications for LBC and will stringently argue the case for greater scrutiny where major change is proposed.

The criteria for considering applications for listed building consent

The publication of PPG 15 marked a subtle but significant shift in policy for the consideration of applications for LBC. Paragraph 3.4 states that applicants *must* be able to justify their proposals. Thus applicants are placed under a duty to submit the proposals, bearing in mind certain relevant policy issues. Separate criteria have been established for assessing applications to demolish a listed building, as distinguished from the general criteria (para. 3.5) for assessing applications for LBC.

The general criteria are listed below.

1. *The importance of the building, its intrinsic architectural and historic interest and rarity, in both national and local terms.*

The grading of a building is a material consideration in determining whether to grant LBC. Yet while grades I and II* buildings are considered to be of outstanding quality, this does not mean that grade II buildings should be regarded as fair game for developers as they make such an important contribution to the wider historic environment.

2. *The particular physical features of the building (which may include its design, plan, materials or location) which justify its inclusion in the list: list descriptions may draw*

attention to features of particular interest or value, but they are not exhaustive and other features of importance (e.g. interiors) may come to light after the building's inclusion in the list.

Thus the special interest which is required for a building to be listed has, as mentioned earlier, a wide meaning including the possibility of finding further matters of interest after a building has been listed, such as the discovery of hidden wall frescos. Therefore, there are no common rules to follow as each building will have its own individual special interest which must be assessed to see if the character of the building will be significantly altered through proposals sought via an application for LBC.

In fact it is recognised in para. 3.13 that it may be possible, in many cases, for sensitive alterations to be made to listed buildings, particularly where a new economic use can be achieved. This, in turn, may help to ensure the objective of long-term preservation. To achieve this certain principles of conservation philosophy based on such concepts as minimum intervention, honesty in repair and generally respecting the integrity of the historic fabric (Brereton, 1991) should be followed (see Chapter 5). Thus a would-be applicant for LBC must devise a defensible change and design solution in order to justify any alterations which are proposed. In this respect anyone considering undertaking works to a listed building should seek expert advice. Moreover, official advice on alterations to external elevations, roofs, external doors, windows, shop fronts, interiors, floors, minor additions and new services is available to guide applicants in the right direction (Annex C, PPG 15). The end result should be to strike an appropriate balance between the special interest in a listed building and change. Accordingly, this will require flexibility on both sides.

3. *The building's setting and its contribution to the local scene, which may be very important, e.g. where it forms an element in a group or townscape, or where it shares particular architectural forms or details with other buildings nearby.*

While s. 16 and s. 66 of the P(LBCA) Act 1990 require LPAs, in considering applications for planning permission or LBC, to have regard to the desirability of preserving the setting of a listed building, the inclusion of 'the setting' as a criterion for judging LBC applications for the first time under PPG 15 has given this issue greater prominence. This would appear to be a recognition of the benefit to community interests which listed buildings provide, not just in the buildings individually but also in relation to the townscape harmony created by particular groupings of buildings and the spaces between and around them.

Moreover, in the case of *Worsted Investments Ltd v. Secretary of State for the Environment and Another* [1994] EGCS 66, it was determined that a planning inspector, in agreeing with the LPA's decision to refuse planning permission, had correctly balanced consideration of a proposal for new development in the setting of the listed Tremall Priory, which would have funded the restoration of the priory, against the damage that would have occurred to the setting. Another factor here was that little of the original building remained but the remains were nevertheless important to the character of the area and doubt was expressed regarding the restoration scheme itself in terms of how much of the original building would end up as being new work. (See also T/APP/P3230/A/92/212334/p7 [1993] JPL 797.)

4. *The extent to which the proposed works would bring substantial benefits for the community, in particular by contributing to the economic regeneration of the area or the enhancement of its environment (including other listed buildings).*

This positive criterion is directed at the many listed buildings considered to be at risk through vacancy, disrepair, or obsolescence. By bringing a building back into use through alteration or repair it can provide an economic use and therefore a benefit to the community. Furthermore, there are many towns and cities throughout the country with buildings 'at risk', often in groups or in a particular area, which are now becoming the subject of regeneration strategies. The Calder Inheritance Project, encompassing the industrial Pennine town of Halifax and its environs, has provided an exemplar to other areas of the benefits of conservation-based regeneration strategies in which all sectors of the community came together to save a large core of historic buildings which had suffered as a result of the decline of the textile industry. The town has prospered economically by re-using and sensitively adapting a range of buildings. Other similar strategies focusing on threatened listed buildings are being developed elsewhere (see Chapter 4).

This new positive criterion may well herald a new era for positive conservation policies. It will require considerable flexiblity by the controlling authorities which is now encouraged by government policy. Further support has been provided by spectators campaigning for 'commonsense conservation' (Catt, 1993). Moreover, research sponsored by the Royal Institution of Chartered Surveyors (RICS) and English Heritage has indicated that many historic buildings have economic potential and have often outperformed other property in investment terms. (RICS/English Heritage/IPD, 1993). It is clear that many listed buildings have a prestige value which has frequently been overlooked by property investors due to the greater risk and uncertainty involved in re-using listed buildings which do not always meet perceived institutional investment requirements.

Thus PPG 15 provides a more positive regime for the conservation of listed buildings than its policy predecessor. Nevertheless, while flexibility and imaginative solutions may be possible, applicants for LBC must be fully aware of the need to respect the integrity of listed buildings.

Cases involving alterations

In making a proposal to alter a listed building the applicant must justify what is proposed to the decision-making body (normally district and borough councils as LPA). Examples of works which have been ruled to have a detrimental effect upon the character of a listed building include such wide-ranging matters as the replacement of Victorian stained glass with clear glazing (APP/Q9495/F/86/120 [1986] JPL 194), the removal of a partition and other internal alterations in a public house (APP/5250/E/82/147 [1983] JPL 751), the replacement of the 'TELEPHONE' lettering on a type K6 listed telephone box with the word 'PHONECARD' ([1989] JPL 810), and the removal of purported chattels as indicated in Chapter 1.

Numerous different examples can be pointed out but it may be useful to indicate some types of alterations which are frequently the source of dispute.

The painting of listed buildings

Since the case of *Windsor and Maidenhead Borough Council v. Secretary of State for the Environment* [1988] JPL 410, which involved two houses externally repainted in deep pink with black detailing within a terrace of matching white finish, it became government policy (as indicated in an amended para. 93 of DoE Circular 8/87 – now see Annex C of PPG 15) to consider whether repainting would affect the character of a listed building (Fig. 2.1). The effect of this can be seen in a case reported in 1990 in which the Secretary of State dismissed an appeal against an enforcement notice as the alteration failed to preserve the character of the building and should have required LBC (APP/F/88/E3715/2–6 [1990] JPL 223). Here the woodwork on the ground-floor window, door frame, fanlight and hood on the front elevation of a Georgian building had been repainted in black.

The alteration of windows

This is probably the most common threat to the character of historic buildings today whether listed or not. Moreover, the subject has become the focus of an English Heritage campaign to educate builders and owners of buildings against carrying out piecemeal repairs by using the wrong materials (Fidler, 1991a, b).

The use of uPVC replacement windows has led to a number of appeal cases. A representative example may be indicated in a case from the Lake District in which the Special Planning Board took listed building enforcement action against the insertion of uPVC 'sash' style windows in Glenridding House (APP/Q9495/F/86/120 [1988] JPL 194). The objective had been to provide double glazing. The change was found to be detrimental to the character of the house which was located in a prominent

Fig. 2.1 The painting of external stonework will usually require listed building consent. The treatment here uses a stabiliser prior to repainting with a sandstone coloured masonry paint to match the upper floors necessary because the original stonework had been repaired previously using red and yellow sandstone replacements and renderings. The alternative of providing new inserts was rejected on cost grounds although this would have produced a more satisfactory result in conservation terms.

position and considered to be of particular architectural interest within the National Park. The owners were required to restore the windows to their former appearance by the insertion of replicas constructed of timber.

In the early 1990s the problem of unauthorised window replacement had become so acute that the Secretary of State for the Environment requested LPAs to have a more robust and committed attitude to resolving it. Of significance, the evidence of action against a local authority serves as a useful indication of the government's view on the question of uPVC replacements. The Secretary of State refused to grant retrospective LBC to the London Borough of Bromley for the installation of uPVC windows in Beckenham Public Hall, a grade II listed building owned by the council. The bad publicity served as an example to other authorities as, at about the same time as making its own application, the council refused to give a private applicant LBC for uPVC windows in his own home (Calvocoressi, 1990).

The main theme of government advice is that the importance of 'appropriate' window design should be recognised. Thus replacements in acrylic-finished aluminium without glazing bars have been found to alter the character of listed buildings (T/APP/N/1405/90/E/806631/2/p7 [1991] JPL 1169). However, in 'special circumstances' the policy criteria on windows may be relaxed. This can be seen in the case of *South Lakeland D.C. v. Secretary of State for the Environment and Rowbotham* [1991] JPL 440. Here the occupier of a listed terraced house had replaced the original sash windows with casement type windows. The occupier was 84 years old, frail and had been unable to open and close the sash windows, and argued that double glazing was required to shut out noise and conserve heat. In a rare decision, on what seems to be compassionate grounds, the court dismissed the action. However, the judgment makes clear that no precedent was meant to be created by this decision. Thus it is unlikely to open the floodgates to similar decisions in the future.

Changes of use

It is officially recognised that the best use for a listed building will often be the use for which it was originally designed. Moreover, the continuation or reinstatement of that use should be the first option (Fig. 2.2). However, para. 2.18 of PPG 15 indicates that new uses for listed buildings may often be the key to their preservation, particularly where a building has become functionally obsolescent. The policy advice in this area is, however, less strict than its predecessor, DoE Circular 8/87, which emphasised the need to find new uses which would not damage the historic fabric of buildings.

Flexibility is now the key word to ensure a building's survival. For example, redundant historic farm buildings may more easily lend themselves to adaption for a business or light industrial function in order to provide a new economic use (Field, 1993) rather than conversion for residential purposes

Fig. 2.2 Heatherslaw Mill, Northumberland. An eighteenth-century grade II listed corn mill, extended and remodelled *c.* 1830 and now preserved in its original use, albeit as a café and working museum retaining all the nineteeth-century machinery.

with the consequent need for new window and door openings and changes in roof elevations (Figs 2.3 and 2.4). Yet, in practical terms, a residential use for such buildings may be the only available economic use, particularly in remote locations, and with a sensitive design much of the historic character of a building can be retained (Morgan, 1991). Therefore, while new uses are to be encouraged, sensitive design solutions need to be worked out so that as much as possible of the special interest in a particular building may be retained. This shift in policy will hopefully resolve the need for buildings to be de-listed following alteration works as – taking again the example of farm buildings – has happened to a number of barn conversions to residential premises throughout the 1980s (English Heritage, 1990).

This more positive attitude was reflected in the case of *Lambeth L.B.C. v. Secretary of State for the Environment and London Residuary Body* [1992] EGCS 17. Here Lambeth L.B.C.'s appeal to the House of Lords against the decision to allow LBC for the conversion of the former County Hall into a hotel was dismissed. The most appropriate use for the building would be for local government, but there was no possibility of the use being resumed. The Secretary of State's decision to grant LBC had been correct as the proposal for a hotel use properly respected the features of the building.

This reflects another aspect concerning proposals for the change of use of listed buildings. Under s. 66(1) of the P(LBCA) Act 1990 a duty is imposed on the decision maker in considering whether to grant planning permission for development which affects a listed building (or its setting) to have special regard to the desirability of preserving the building or its setting or any features of special architectural or historic interest which it possesses. In *Heatherington (UK) Ltd v. Secretary of State for the Environment and Westminster City Council* [1995] JPL 228, the Deputy Judge confirmed that the objective of preservation had 'considerable weight'. Thus while the statutory duty to have regard to the policies of the development plan under s. 54A of the principal planning Act is important, the material consideration 'to have special regard to the desirability of preserving' may nevertheless be first in importance for the decision maker in considering *any* scheme for alternative use. The case had arisen due to the fact that, at appeal, the planning inspector had rejected a proposal to allow the continuation of office use, which had had temporary permission for office use, while not taking the same approach regarding the LPA's proposed alternative scheme for residential use which was favoured by the adopted development plan policy (Edwards and Martin, 1995).

A useful case example to consider the potential problems in accommodating new uses can be indicated from an appeal case concerning Clarence House, a large listed building in the main shopping area of Brighton (T/APP/N/1405/90/E/806631/2/p7 [1991] JPL 1169). It was originally built as a coaching inn and had been used as an hotel. However, in recent years most of the interior had been used as offices. The appeal arose over part of the ground floor which had been isolated from the office accommodation and for which a proposal had been made for conversion to a shop.

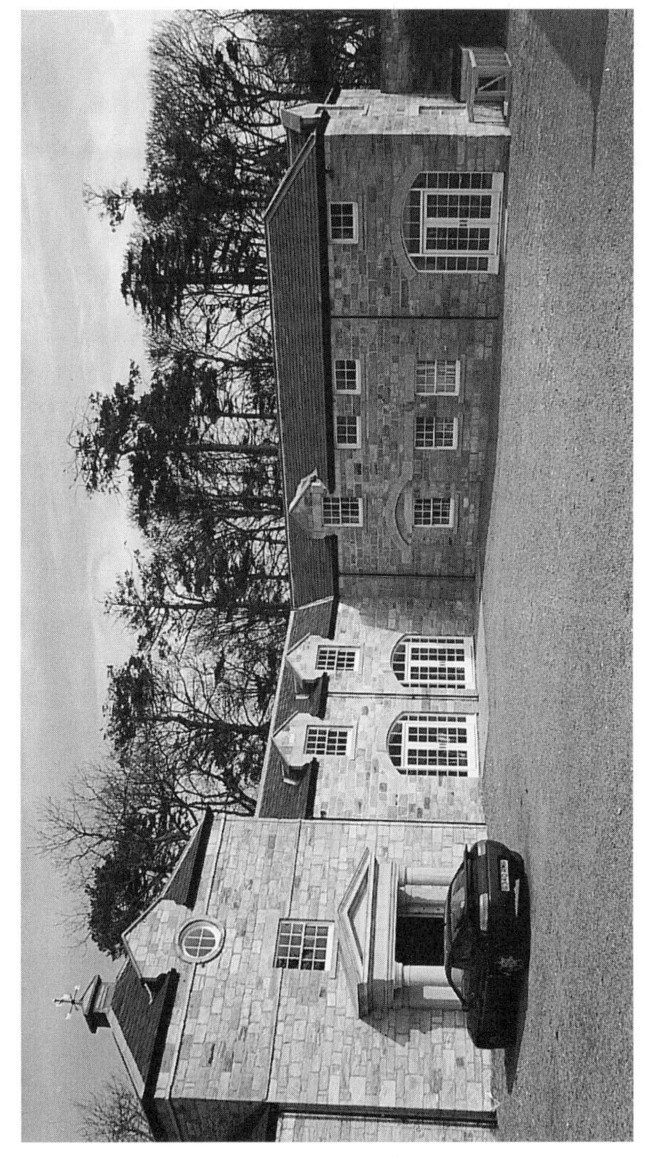

Fig. 2.3 Listed estate farm buildings at Felton, Northumberland. These barely resemble their original use having been substantially altered for residential use with window openings piercing the roof line and the addition of a classical portico entrance.

Fig. 2.4 Listed farm buildings in a North Northumberland hamlet. Converted to residential use with damaging alterations including a stone chimney, metal flue, velux roof windows, and a large conservatory extension.

The appellants, who had sought the specialist advice of a reknowned architectural historian, successfully argued that the changes would not be detrimental to the character of the building and, more significantly, would enrich the building.

The inference here is that alterations, if handled skilfully (and in line with policy criteria), can enhance the character of a listed building. Moreover, cumulative changes over the life of a building may be of special interest. However, it is necessary to point out that the proposal here had not affected the main façade, the most significant aspect of architectural interest in the building, but related to a blank side elevation which once had openings.

The opposite view is that proposed conversions which are insensitive to the special interest of a listed building should not gain LBC. A good example to illustrate this point can be indicated in relation to a proposal to convert the grade II listed Stotfold Mill, Bedfordshire. The application for LBC was called-in by the Secretary of State, as it involved 'works of demolition' in that a substantial part of the interior mill machinery was proposed to be removed, and considered at a Public Inquiry in 1986 (Ref. E1/5124/270/71 and E1/5124/411/4). LBC was refused for the removal of mill machinery that was not actually fixed to the listed building but was nevertheless found to form part of the special interest in the building. This then thwarted the proposal to convert the mill to three residential units.

Another similar example may be indicated from the case of *Cotswold D.C. v. Dogdaisey Ltd (Tetbury)* (unreported 1987). Here the Magistrates' Court fined the owners for removing machinery from the mill building without consent and ordered their repurchase and return which, in turn, prevented the proposed change of use (Carter *et al.*, 1992).

Where a building is so sensitive that it is not able to sustain a new economic use there are two alternative possible courses of action which may be considered. The first is to secure it by charitable or community ownership. For example, a redundant ecclesiastical building may be secured by the Churches Conservation Trust (formerly the Redundant Churches Fund). The National Trust or one of the many Building Preservation Trusts may also provide guardianship. The other alternative is to consider the merits of allowing a listed building to be demolished.

Demolition works

Leaving aside the problems associated with the actual definition of demolition, para. 3.17 of PPG 15 indicates that the Secretary of State would not normally expect consent to be granted for the total or substantial demolition of any listed building. In order to obtain consent an owner must be able to provide convincing evidence that every effort has been made to continue the present use or find another alternative use. Moreover, the demolition of a grade I or grade II* building requires 'the strongest justification'. However,

PPG 15 signified the removal of the strict *presumption in favour of preservation* in that if there is clear evidence that redevelopment could produce a 'substantial planning benefit for the community' which outweighs the loss arising from demolition, LBC may be granted.

Paragraph 3.19 of PPG 15 specifies three policy criteria by which applications to demolish listed buildings must be considered:

1. *The condition of the building, the cost of repairing and maintaining it in relation its importance and to the value derived from its continued use. Any such assessment should be based on consistent and long-term assumptions. Less favourable levels of rents and yields cannot automatically be assumed for historic buildings. Also, they may offer proven technical performance, physical attractiveness and functional spaces that, in an age of rapid change, may outlast short-lived and inflexible technical specifications that have sometimes shaped new developments. Any assessment should also take account of the possibility of tax allowances and exemptions and of grants from public or charitable sources. In the rare cases where it is clear that a building has been deliberately neglected in the hope of obtaining consent for demolition, less weight should be given to the costs of repair.*

This criterion significantly differs from its predecessor contained in para. 90(c) of DoE Circular 8/87, which recognised that 'old buildings generally suffer from some defects', by emphasising the positive values which historic buildings may have to offer over modern equivalents. The investment potential of listed buildings realised here, which has been confirmed by research findings (RICS/English Heritage/IPD, 1993), is a further example of changing perspectives on the value of our built heritage. Nevertheless, while the updated test is more positive, the common law rules which have applied to proposals seeking demolition on the question of condition and economic viability are still relevant.

Applications for the demolition of listed buildings are perhaps most frequently based on the premise that the costs of preservation make the project economically unviable. Moreover, it is not just the costs of preservation which are of significance but also whether, once repaired, a building can be put to an economic use. The courts have established that if the repair of an unused listed building does not provide an economic proposition for a new use it is appropriate to consider, against all other relevant matters, the relevancy of a proposed replacement building in this respect. In *Kent Messenger Ltd v. Secretary of State for the Environment* [1976] JPL 372 it was stated that the following matters will be of significance:

- the cost of putting the listed building into good repair;
- the value of the building for any purpose if put into good repair;
- accordingly, the extent to which such restoration is an economic proposition;
- whether any, and if so what, replacement building is feasible; and
- the cost and value of any such replacement.

Philips J. in the *Kent* case quashed the Secretary of State's refusal to allow demolition because he had failed to take into account the argument that not only would the building be costly to repair but also that, once repaired, the building would not be an economic proposition for a new use. A similar outcome resulted in the case of *Trustees of the Bristol Meeting Room Trust v. Secretary of State for the Environment and Bristol City Council* [1991] JPL 152, as the Secretary of State misinterpreted his inspector's finding that 'he was not satisfied that the building was incapable of economic restoration' to mean that the building was actually incapable of economic restoration. Had he endorsed the inspector's view that, on the evidence presented, he was unconvinced that the building was capable of economic restoration, this would have sufficed. At the same time the specific reference to the fact that less weight will be given to repair costs in cases of 'deliberate neglect' marks a much stricter policy direction for the consideration of relevant evidence.

However, while the Secretary of State has a duty to direct himself in relation to established policy, an applicant must also take account of the need to produce actual evidence as to the cost of repair/adaption works (including evidence that avenues for financial assistance, such as the possibility of a repair grant from English Heritage, have been investigated), and of the resulting value of the building. There are many examples of cases which have failed because the economic argument has not been fully investigated. For example, in *Henry Davies & Co. v. Secretary of State for the Environment and Southwark L.B.C.* [1992] EGCS 61, 'allegations' as to the condition of a listed building and the viabilty of a new use did not provide adequate evidence, as compared to actual costings, to allow the building to be demolished. Moreover, ignoring the possibility of the availability of a repair grant in calculations concerning economic viability may be sufficient to defeat the case for granting LBC (see SW/P/5117/147/8 and APP/U0110/A/88/093973 [1992] JPL 1186). Furthermore, the fact that policy guidance states that 'less favourable levels of rents and yields cannot automatically be assumed for historic buildings' means that applicants must provide clear evidence that a building has no value in real terms for its original or a potential new use. This issue is further considered in the second criterion:

2. *The adequacy of efforts made to retain the building in use. The Secretaries of State would not expect listed building consent to be granted for demolition unless the authority (or where the appropriate Secretary of State himself) is satisfied that real efforts have been made without success to continue the present use or to find compatible alternative uses for the building. This should include the offer of the unrestricted freehold of the building on the open market at a realistic price reflecting the building's condition (the offer of a lease only, or the imposition of restrictive covenants, would not normally reduce the chances of finding a new use for the building).*

In this context, para. 89 of DoE Circular 8/87 simply stated that the Secretary of State 'would normally require to see evidence that the freehold of the building had been offered for sale'. Thus, the new criterion provides a much stricter marketing test for demolition applications. Furthermore, from both criteria 1 and 2 it is clear that the policy regime under which proposals to demolish listed buildings are to be considered following the publication of PPG 15 will necessitate much greater scrutiny of relevant evidence produced by applicants.

It should also be remembered that while economic viability arguments probably provide the strongest case for allowing demolition, the more general criteria for assessing applications for LBC also have relevance. Despite the loss of the strict 'presumption in favour of preservation', the *Bristol Meeting Room* case established that paramountcy must be given to the duty contained in s. 16(2) of the P(LBCA) Act 1990 of having 'special regard to the desirability of preserving a listed building' in determining an application proposing demolition. Thus, in *Cullimore v. Secretary of State for the Environment* [1992] EGCS 69, cost was found to be only one of the factors at issue – the intrinsic historical interest and rarity, as now expressed in para. 3.5(i) of PPG 15, was of sufficient importance for not allowing the remains of a listed farmhouse to be demolished.

The third criterion allows scope to justify the demolition of a listed building to allow for new development:

3. *The merits of alternative proposals for the site. While these are a material consideration, the Secretaries of State take the view that subjective claims for the architectural merits of proposed replacement buildings should not in themselves be held to justify the demolition of any listed building. There may very exceptionally be cases where the proposed works would bring substantial benefits for the community, which have to be weighed against the arguments in favour of preservation. Even here, it will often be feasible to incorporate listed buildings within new development, and this option should be carefully considered: the challenge presented by retaining listed buildings can be a stimulus to imaginative new design to accommodate them.*

Before explaining the purpose of this criterion it is perhaps useful to consider how conservation policy has changed as regards the merits of allowing alternative proposals for a listed building site.

Although 'preservation' was stated to be a 'paramount consideration' in the *Bristol Meeting Room* case it does not necessarily override other considerations which may be relevant in planning policy. Thus, in a few 'exceptional' circumstances demolition has been permitted in situations where economic reasons have not prevailed. For instance, in 1988 the Secretary of State for the Environment granted LBC for the demolition of a grade II listed building in order to allow investigation of the remains of a Roman Amphitheatre under the building (DNW/5143/344/4). This was one of only three such legionary fortresses in Britain and therefore of national importance. The decision reflected the growing importance of archaeological matters in the sphere of town and country planning which, with the discovery of other important archaeological remains in the late 1980s, subsequently led to the publication of a planning policy guidance note on 'Archaeology and Development' (see Chapter 8).

More specifically, the former policy criterion in para. 90(d) of DoE Circular 8/87 indicated that LBC for demolition could be given to allow new development which could enhance the environment of a rundown area, particularly where this action would be for some public purpose or would benefit other listed buildings. Thus, in *R. v. Westminster City Council, ex parte Monahan* [1989] JPL 107, it was confirmed that the raising of finance as a 'planning gain' for the restoration of the principal parts of the grade I listed Royal Opera House at Covent Garden could be justified by the demolition of other listed buildings of less importance in order to carry out an otherwise unacceptable office development (insufficient funds could be raised for the restoration by other means).

Yet this is a rare example and the possibility of an alternative use of the site generally has not outweighed the paramountcy of preservation. Moreover, the economic viability of a listed building may turn upon the question of area improvement. This can be seen in the case of *Godden v. Secretary of State for the Environment* [1988] JPL 99. Here the appellant argued that the cost of repairing the property (a listed former seaside café) and the value of the property when restored would have left a shortfall of £60,000. It was also contended that in examining the question of economic value only the building alone could be considered and not the site. Stuart Smith J. accepted that the potential for redevelopment was a material matter which could be considered by the Secretary of State. He further felt that the weight to be attached to this material consideration would depend on how fixed and certain proposals were for the site. A development brief had been prepared for the area and a rival scheme had been put forward which included the listed building. Thus it was entirely reasonable to take into account this alternative proposal for the site and the decision to refuse demolition was upheld.

The most significant 'exceptional' circumstance decision may be indicated in the case of *Save Britain's Heritage v. Secretary of State for the Environment and Others* (1991) LG Rev 429. In the now infamous saga of the Mappin and Webb site at No. 1 Poultry, Mansion House Square, London, the House of Lords confirmed the Secretary of State's decision to grant LBC for the demolition of eight listed buildings because of the overriding design merits of the proposed replacement building by the now deceased architect James Stirling, despite the fact that the Prince of Wales had likened the design to a 'wireless set'. The campaigning body, Save Britain's Heritage, considered that the merits of refurbishing the grade II listed buildings which had 'group value' had not been properly considered. Nevertheless, their Lordships confirmed that the design of a replacement may be a relevant consideration in determining whether to grant LBC. However, they were careful to point out that this decision was not meant to act as a precedent, thus maintaining the concept of preservation as a paramount consideration. It was only the unique location of appeal site in a unique conservation area characterised by other listed buildings of the greatest architectural distinction that led them to conclude that the replacement building 'would be a worthy modern addition to the architectural fabric of this part of the City'. It was further stated to be unlikely that the circumstances of the case could ever be repeated.

Despite this, the positive wording of criterion 3 and reference in para. 3.17 of PPG 15 recognise the possibility of other similar outcomes due to the countervailing value in planning terms of development proposals which may merit permission on the grounds of public benefit. This is in stark contrast to the negative guidance of DoE Circular 8/87 which emphasised the intrinsic value of the listed building without indicating clear reasons for the justification of demolition. Therefore, while demolition will not normally be permitted because a redevelopment is more economically attractive, 'where a substantial benefits for the community would decisively outweigh the loss resulting from demolition' it may be acceptable. Moreover, in the case of a proposal to demolish a listed building which lies within a conservation area, the case of *Davis v. Secretary of State for the Environment and Southwark L.B.C.* [1992] JPL 1162 has confirmed that the relevant statutory provisions are interdependent in that a 'suitable redevelopment scheme' must meet the requirements for granting LBC for demolition and planning permission for new development under both protection regimes.

The meaning of 'substantial community benefit' is not entirely clear and will no doubt be revealed in the course of time through case decisions. In the meantime there are a number of issues which may be suggested as fitting this requirement. For instance, a community benefit may take the form of a LPA or Urban Development Corporation strategy for economic regeneration, particularly where historic buildings can offer a focus for regeneration. The 'Grainger Town' regeneration scheme in Newcastle upon Tyne is such an example (see Chapters 4 and 7).

Development schemes of architectural excellence may be another form of planning benefit which could justify demolition. Thus criticisms of Lord Palumbo's proposals for the No. 1 Poultry site were outweighed by the design of the proposed replacement building regarded by others as a 'potential masterpiece' and, as such, would be of greater aesthetic value to the community than the existing listed buildings.

Another potential planning benefit may be indicated to include 'enabling development' which allows another listed building to be repaired or restored as in the *Monahan* case involving the Royal Opera House. The meaning of the word 'community' is more difficult to determine. For instance, there may be situations where the detriment to the local community caused by demolition may be outweighed by the benefit to the national community of an internationally praised replacement design (Millichap, 1993).

Façadism

The question of façadism, i.e. the partial demolition of listed buildings behind their façades and redevelopment of a new building, has been the subject of critical debate for many years. The pure conservation view is that façade retention creates a sham, while others frequently argue that partial demolition is the only answer to secure economic preservation, albeit of the front, of a listed building. This may have particular significance in conservation areas where the context of the street scene has recognised importance.

The principal argument against façade retention is that it results in the loss of a building's architectual and historic integrity. Bearing in mind that listing status is afforded to the whole building, the loss of features behind the main façade would appear to be contrary to the basic tenets of the statutory regime of protection. English Heritage has confirmed this view in an advice note on conservation areas as recently as 1993 by highlighting that the character of an area depends upon more than just the street frontages of its buildings. In other words, their integrity as historic structures and the contribution they make in all dimensions, including the roofscape, back elevations and side views, should be considered as relevant factors (English Heritage, 1993). Moreover, PPG 15 signalled a strengthening of policy views by indicating in the last sentence of para. 3.15 that *'the preservation of façades alone, and the gutting of and reconstruction of interiors, is not normally an acceptable approach to the re-use of listed buidings: it can destroy much of the building's special interest and create problems for the long-term stability of the structure'*. Thus, aside from the loss of heritage value, evidence from façade schemes already completed has shown that the necessity for additional support has not always been fully anticipated leading, in some cases, to the collapse of the façade (Pertee, 1991).

While technical problems associated with façade retention may be resolved relatively easily through detailed planning of each stage of the works, opposition to façadism on preservation grounds is not so quickly exhausted. Yet in an attempt to satisfy modern requirements for living and working environments it is often financially necessary to consider total redevelopment of the interior of a listed building. This course of action may well be the only economically feasible method of ensuring the re-use or continued use of certain buildings.

By example, in Grey Street, Newcastle upon Tyne, described by Gladstone as 'our best modern street' in 1862 (Morley, 1903) and today considered to be of significant architectural merit in national terms (Pevsner *et al.*, 1992), approximately half of the listed neo-classical buildings have been subject to redevelopment schemes behind retained façades (Figs 2.5 and 2.6). The street is one of the prime office locations in the city but until the mid-1980s had attracted little investment money due to a long-standing pattern of low rental returns. This had two resulting effects, i.e. deterioration and vacancy, and pressure for redevelopment. On the basis of economic viability

Fig. 2.5 Grey Street, Newcastle upon Tyne. One of the finest architectural streets in England where many of the buildings have been redeveloped behind retained façades while others lie in wait for a similar fate with empty upper floors. Grey Street forms part of the Grainger Town Conservation Area Partnership area, one of the first 14 pilot project areas under this new conservation funding arrangement.

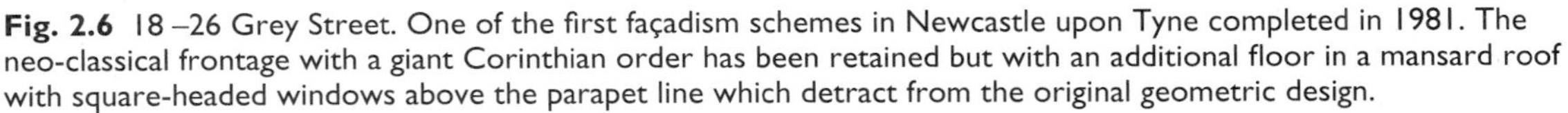

Fig. 2.6 18–26 Grey Street. One of the first façadism schemes in Newcastle upon Tyne completed in 1981. The neo-classical frontage with a giant Corinthian order has been retained but with an additional floor in a mansard roof with square-headed windows above the parapet line which detract from the original geometric design.

arguments the LPA allowed a number of façadism schemes involving the wholesale removal of interiors, the creation of additional floors within the original building height but at variance to the fenestration and by mansard roofs above the original parapet line.

The fortunes of Grey Street started to improve in investment terms, allowing, as in other similar examples around the country (Catt, 1987), the provision of modern office accommodation and the retention of a harmonious historic streetscape, one of the policy objectives behind the City Conservation Area in which Grey Street lies. Furthermore, in 1988 the Grey Street Renaissance initiative was devised to further stimulate interest and investment while maintaining the architectural character of the area. This façadism approach to conservation, apparently favouring area protection before individual properties, seemingly won support from English Heritage (Newcastle City Council *et al.*, 1988)

This view must be tempered by the fact that the merits of each conversion proposal are individually assessed and that others have expressed concern that the pace of façadism approvals has transformed Newcastle's townscape to nothing more than a 'stage set' (Henderson, 1989). Moreover, the initiative has indicated the possibility of pursuing repair action against reluctant owners of decaying property in order to maintain the interiors of recognised merit. Yet due to the practical problems associated with such action (see Chapter 4) this has not happened, fuelling the argument for façadism proposals. In particular the case of one complex of buildings, the nineteenth-century façade housing an earlier eighteenth-century interior, vacancy and lack of maintenance for some 20 years has irretrievably damaged the special interest of the interior, and despite attempts to scale down the owner's redevelopment proposals economics will eventually dictate that a partial demolition scheme be allowed.

This case example is not singular to Newcastle and therefore shows the limitations of policy desires as expressed in PPG 15 that the preservation of façades alone is normally to be regarded as unacceptable. Moreover, while listed status applies to the whole building, the requirement to obtain LBC only applies where works affect the character of a building, or, in other words, where its special interest will be prejudiced. Accordingly, when interiors have little or no architectural or historic merit, which is often the case in buildings listed for 'group value', the argument against façadism redevelopment schemes is difficult to prove (Millichap, 1994). This is even more so when existing interiors are in a dilapidated condition and beyond economic repair, or where existing internal elements are structurally incapable of supporting the loads required to find a new and viable use (Highfield, 1991).

Furthermore, as in the case of Grey Street, the principal reason for widespread acceptance façadism is the considerable value placed upon the preservation of townscape. This would seem to fall within the criteria of 'substantial planning benefit for the community' envisaged in PPG 15, as is supported by conservation area legislative provisions. Thus in an appeal relating to a grade II listed building in Hanover Square, London, while it was found to be economically possible to repair and refurbish the original structure, development behind the façade would create a development with a long life expectancy and to modern standards (APP/X5990/A/87/076725). Although the LPA's retention proposals were less expensive, it was concluded that they were unlikely to be cost effective in the short term and not justified commercially in the long term. In areas designated for regeneration policies, this approach may bring long-term community benefits to an area.

Dismantled and naturally destroyed listed buildings

Can a listed building cease to be a building and thereby lose its protection? This is an important question to which the answer in simple terms is that a building will cease to be listed when the Secretary of State for National Heritage decides to remove it from the statutory list. Buildings that have been neglected can be de-listed if the special architectural or historic interest has been lost or reduced. However, if a listed building has been the subject of unauthorised alteration works or demolished without LBC, a listed building enforcement notice (LBEN) may be served under s. 38 P(LBCA) Act 1990 specifying the steps required to restore the building to its former state, or otherwise alleviate the effect of works carried out without LBC. But when a building has been dismantled without LBC, it is not clear whether the building still exists or whether it is relevant to insist that listed building protection laws continue to apply.

This issue was raised in the case of *R. v. Leominster District Council, ex parte Antique Country Buildings Ltd and Others* [1988] 2 PLR 23. Here a listed barn had been dismantled so that it could be exported. Mann J. decided that a demolished listed building was none the less still a 'building'. He also found that enforcement action for restoration could be pursued if the components of a building are extant. It would be a matter of fact and degree, in this case 70 to 80 per cent of the structural timbers were extant and thus the LPA were entitled to issue a LBEN requiring restoration of the barn. The dismantling amounted to a breach of listed building control as it involved demolition works without LBC.

The *Leominster* case was followed by a similar appeal case (APP/L2630/A/87/079563). The circumstances were slightly different as the building had not been dismantled for the purposes of re-erection elsewhere but rather because it had partially collapsed after a storm. The building was a barn of significant historical, architectural and landscape value. Whether or not the timbers in this case constituted a 'building', the majority parts of the frame existed. The inspector felt that, with some repair and replacement, the frame could be re-erected as enough of the structure remained with sufficient joiners' marks for this to be undertaken with reasonable accuracy.

However, the main point concerning this appeal was whether a new use should be allowed for the re-erected barn. It was argued that this would allow the preservation of its special qualities and that the preservation of a listed building should override the structure plan policy against new residential redevelopment (the proposed new use). Having accepted on the basis of the *Leominster* case that the building still existed, it was considered on balance that preservation was, in effect, the paramount consideration. Planning permission was granted for the re-erection of the barn with a new use subject to conditions concerning the external design.

While in this instance only planning permission was required, and not LBC, the barn had been granted planning permission and listed building consent for an earlier proposal, before the barn had collapsed, which was virtually identical to the scheme in question. This is nevertheless at variance to the *Leominster* case where it was found that the dismantling of the building could be subject to a LBEN. The main issue here, however, was whether the building could be ordered to be re-erected on its original site. The two cases can also be differentiated as it is unlikely that a LPA would serve a LBEN (nor is the notice likely to be upheld) where the damage to a listed building was not wilful, but, for example, by storm damage or fire (Tromans and Turrall-Clarke, 1991). (See also *Burhill Estates Ltd v. Woking Borough Council* [1995] JPL 147.)

Thus a dismantled listed building will remain a listed building (and subject to planning and listed building control) if the majority of the materials (i.e. over 50 per cent) are still extant and it is possible to re-erect the building with some degree of accuracy (albeit with some new materials being used). Where a building has not been purposely dismantled the only possible course of action to ensure its re-erection would appear to be via the serving of a Repairs Notice under s. 48 of the P(LBCA) Act 1990.

The setting of a listed building

The inclusion of 'the setting' as a criterion for judging whether LBC should be granted under para. 3.5(iii) of PPG 15 will have particular relevance in situations where listed buildings are the focus of and integral to a regeneration or redevelopment scheme (Fig. 2.7). As in the *Davis* case, the grant of LBC and planning permission are likely to be interdependent. In other respects, individual development proposals, isolated from works to listed buildings, but nevertheless likely to affect the setting of a listed building, have already been subject to particular scrutiny. A number of appeal and court decisions provide evidence of the policy desire now expressed in paras 2.16 and 2.17 of PPG 15 to safeguard 'the setting' in relation to the townscape value of attractive historic streets, village locations and individually in the context of a building itself (Fig. 2.8).

In a 1985 appeal case involving a listed building in Selby (APP/B2735/

Fig. 2.7 Formerly a maternity hospital c. 1826 and more recently the BBC Broadcasting House in Newcastle upon Tyne which has become the focus of a redevelopment scheme. The surrounding development replaces a mixture of brick buildings in poor condition. The scheme resulted in the removal of a 1960s addition to this listed building, which detracted from its character, and improved its setting by providing new railings and street improvements as well as a background to the building which attempts to harmonise with, and offer a contribution to, the existing townscape.

Fig. 2.8 The John George Joicey Museum. One of only two complete 17th-century brick buildings to survive in Newcastle upon Tyne, now engulfed by the 1960's British Telecommunications' building Swan House and the inner city transport infrastructure in the 1960s. Today listed buildings are given greater scrutiny in redevelopment schemes.

E/800378), the inspector ruled against its conversion into four dwellings, as it would have been detrimental to the 'expansive quality befitting a grand domestic house', and the erection of three dwellings within its grounds which would have impinged upon its setting.

A similar conclusion was drawn in relation to the individual setting of a complex of grade I, II* and II buildings, namely Bollitree Castle near Ross-on-Wye, in the case of *R. v. South Herefordshire District Council, ex parte Felton* [1991] JPL 663. The owner successfully brought a case of judicial review to quash the decision of a LPA to grant planning permission for a potato store on the adjoining farm, even though the store had been erected, on the basis that it had not properly considered the effect it would have had on the setting of the listed buildings. Despite the fact that the store was not visible from within the main house, except by an attic window, it was visible from the approaches to it and generally dominated the listed buildings.

Three cases may be indicated regarding village settings of listed buildings. The first of these concerned a proposal to erect 14 detached houses and garages close to a grade II listed church, the scheme was regarded by the LPA as an overintensification of the site as regards the setting of the church, bearing mind that an earlier scheme which had planning permission for nine houses was at a greater distance from the church (T/APP/P3420/A/88/098142/P2). It was argued that the church's listing was largely as a result of its interior decoration and monuments. But, whatever the relative architectural or historic merits of the church, it was the setting of the church that was the important issue. The appeal site was located right at the heart of the village which was viewed by local residents as a place of special importance, especially as it enjoyed a prominent position. It was clear from representations made by local residents that the immediate vicinity of the church was particularly sensitive as it provided the village with its identity. For this reason the scheme for 14 houses was rejected on appeal.

In another case involving a village church, listed as grade I, a residential development scheme was rejected on the grounds that the setting of the church would have been seriously harmed, being visible in its entirety from many places and considered to be the most important feature of the village, apart from the fact that the proposal would have detached the church from its open setting (T/APP/E2530/A/89/144872/P7).

In an appeal involving a proposal to develop a hotel and re-use a collection of listed agricultural buildings at the prehistoric site at the village of Avebury, Wiltshire (SW/P/5407/21/74 (1991)), a number of matters were at issue. The proposal was rejected because, apart from the effect upon the conservation area and scheduled monuments and the fact that the site had received world heritage status, the Inspector felt that the scale of the hotel would overwhelm the listed buildings and would be seriously detrimental to their setting, character and appearance, destroying the traditonal relationship of the buildings to each other and the farmyard.

The wider townscape value of historic buildings was considered in the case of *North Wiltshire District Council v. Secretary of State for the Environment* [1991] EGCS 25. Here it was confirmed, by way of reference to an earlier planning refusal, that the erection of a new dwelling house on land adjoining the Coach House, Norton Lodge, in Chippenham would be detrimental to the listed building and the physical aspect of the site being surrounded by other historic buildings.

The importance of the individual setting of listed buildings and the effect upon a conservation area on the edge of the town of Barrow-in-Furness was considered in a 1991 appeal decision (T/APP/W0910/A/90/173319/P7).

Here the parkland setting of the grade II* Abbey House Hotel (and other listed buildings), designed by Edwin Lutyens in 1914, in the appearance of formally laid out gardens would have been affected by a residential development proposal. It would also have affected the Furness Abbey conservation area which had been designated as an Area of Special Landscape Value in an emerging local plan.

The sanctity of the setting of a listed building is clearly revealed from these representative case decisions. However, no clear definition of 'the setting' can be deduced. In other words, each case must be assessed on its particular merits. In this context, para. 2.13 of PPG 15 indicates that LPAs should seek expert professional advice, such as from the Royal Fine Art Commission, before making decisions on the granting or refusal of planning permission which 'might be held to have an adverse impact on any sites or structures of the historic environment'. Nevertheless, many LPAs have been concerned that there is no formal mechanism to consider whether the setting of a listed building may be affected by development proposals (Suddards, 1988). PPG 15 has gone some way to meeting this criticism in that para. 2.17 indicates that where a listed building forms 'an important visual element in a street' it should usually be assumed that any development in the street will be within the setting of the listed building. The guidance further indicates that high or bulky developments some distance away may also affect the setting, including the historic skyline. In other cases the guidance suggests the need for a historical assessment of a building's surrounding including the possibility of publishing a notice to identify the extent of 'the setting'.

In the case of advertisement control the Town and Country Planning (Control of Advertisements) Regulations 1992 allow for the designation of *Areas of Special Control* which PPG 19 (Outside Advertisement Control) implies may be suitable to protect groups of listed buildings from insensitive advertisement where they are located in a small enclave, such as the precincts of a cathedral and neighbouring ecclesiastical buildings or a historic marketplace. More specifically, para. 24 of PPG 19 advises that special care is needed to ensure that any advertisement displayed on, or close to, a listed building should not 'distract from the building's design, historical structure or character and does not spoil or compromise its setting'. Yet in practice the interpretation of the regulations is not entirely clear (Bagnall, 1995).

However, changes in the statutory provisions regarding the making of development control decisions may dispense with the necessity for any formal mechanism regarding development affecting the setting of a listed building. The introduction of s. 54A into the Town and Country Planning Act 1990, via the Planning and Compensation Act 1991, has resulted in the development plan as being the primary criterion by which development decisions must be made (unless other material considerations dictate otherwise). While LBC applications are not legally subject to the provision of s. 54A, development decisions affecting the setting have to be viewed against relevant policies set out in approved development plans. Accordingly, para. 2.8 of PPG 15 advises LPAs to formulate specific conservation policies within local plans in relation to development proposals which would affect the setting of listed buildings. This will eventually allow clearer guidance for developers regarding new development proposals, and for LPAs to judge the merits of such proposals.

At the same time new development should not be considered as a totally negative matter in relation to conservation policy. As PPG 15 emphasises, the potential of new development to have planning benefits may result in the setting being compromised in some situations. The *Monahan* case is such an example in that otherwise unacceptable office development was allowed due to the planning benefits derived from the potential to restore the Covent Garden Opera House through funds provided by the developers.

This approach was followed in a subsequent appeal decision (APP/G/1820/A/88/103529) involving the grade I listed Croome Court in Worcestershire. The building's owner had argued for the development of eight houses in its walled garden to support the building's restoration. The inspector accepted that the proposal constituted the least visible siting for the development but felt that it would represent a new and incongruous element in a historic setting which would diminish the importance of the garden as part of the setting of the listed building. Thus there would have to be special circumstances for allowing the development. These were found as the appellant's costings showed a deficit of £900,000 on the restoration of the listed building without the proposed development, even with grant aid from English Heritage.

Crown land and listed buildings

Crown buildings can be listed under s. 83(1) of the P(LBCA) Act 1990 but have been exempt from the normal controls affecting listed buildings under the Act. Nevertheless, via DoE Circular 18/84 the government provided that Crown bodies will normally operate as if these controls apply.

In 1992 the government announced, via the publication of a DoE consultation paper, an intention to bring the Crown exemption under the system of town and country planning to an end. Once the necessary legislation has been enacted, the Crown will become subject to the provisions of the P(LBCA) Act with the exception of development proposals involving national security, the compulsary purchase provisions and certain enforcement provisions. A declaratory system rather than the specific enforcement procedures will apply to contraventions in the relevant legislation. However, until the Crown exemption is actually removed, the policy outlined in DoE Circular 18/84 will continue to apply.

Ecclesiastical buildings

Special provisions apply to ecclesiastical listed buildings which are retained in use for ecclesiastical purposes. Under s. 60 of the P(LBCA) Act 1990 such buildings are exempt from certain controls. In practice this includes the following:

- the need for LBC (and conservation area consent) for works of partial demolition, alteration and extension;
- building preservation notices;
- spot listing;
- compulsary acquisition (of listed buildings in need of repair);
- execution by a local authority of urgent works for unoccupied listed buildings;
- criminal liability for damaging listed buildings.

Planning permission may be required for material external alterations and extensions and the total demolition of a church will normally require LBC.

The exemption originates from an agreement made in 1913 between the Archbishop of Canterbury and the government in response to assurances as to the Church of England's future good conduct with regard to its buildings designated for worship (i.e. the church itself and not any associated residence of a priest). Following this the Church of England developed its own system for the control of works to its churches. In practice, this means that a *Faculty Jurisdiction* (or licence) provides the relevant permission for material alterations including works or alterations to a church, churchyard or church furnishings and for the total or partial demolition of a church. This permission is issued by the Chancellor of the Diocese in which the church is situated, via the authority of the *Consistory Court* under Church laws and governing procedures presently found in the Faculty Jurisdiction Measure 1964, the Care of Churches and Ecclesiastical Jurisdiction Measure 1991, the Faculty Jurisdiction Rules 1992, and the Faculty Jurisdiction (Injunctions and Restoration Orders) Rules 1992.

In the case of Church of England cathedrals relevant works must be authorised by the Cathedrals Fabric Commission for England under the Care of Cathedrals Measure 1990, the Care of Cathedrals (Supplementary Provisions) Measure 1994 and the Care of Cathedrals Rules 1990.

The case of *Attorney-General (on relation of Bedfordshire County Council) v. Trustees of Howard United Reform Church, Bedford* [1976] AC 363 HL determined that ecclesiastical buildings, other than Church of England buildings, were also covered by the exemption. This meant that an ecclesiastical building included Church of England churches, other Christian churches and certain other Christian buildings – for example, Lord Denning suggested theological colleges and Bishops' palaces.

While the Church of England has had a stringent internal system for controlling works to church buildings, other Christian denominations have been much slower to take the same approach. In 1989 the Roman Catholic Bishops' Conference passed a resolution in relation to existing churches of particular architectural or historic interest. This stated that any alterations must be carried out with sensitivity and, where appropriate, in consultation with English Heritage. The Methodists' response has been to produce a guidance booklet entitled *A Charge To Keep – A Methodist Response to Listed Buildings and Conservation.* This book has stressed the need to consult with LPA conservation officers when changes are planned.

However, in February 1992 a consultation paper was issued to relevant ecclesiastical and conservation bodies, and local authorities, concerning the ecclesiastical exemption. In December 1992 the Secretary of State for National Heritage issued a response to the consultation exercise, the main findings of which are now found in section 8 of PPG 15. The result of this is that exemption is retained for those bodies which subscribe to a *code of practice* and agree to bring their internal arrangements for control of works to listed ecclesiastical buildings (and those in conservation areas) into conformity with the code. The exemption will be withdrawn for bodies which choose not to subscribe to the code. Guidance has been issued to LPAs on the due weight to liturgical requirements in considering applications for consent to internal alterations.

The basic outline of the code of practice, which is set out in para. 8.4 of PPG 15, is as follows:

1. Proposals for relevant works, which would normally require LBC but for the exemption, should be submitted by the local congregation for approval of a body independent of them.
2. The body should have arrangements for obtaining advice from conservation specialists. (Of note, English Heritage produced a policy statement entitled *New Work in Historic Churches* in October 1991.)
3. The decision-making process should provide for:
   - consultation with the relevant LPA, English Heritage and the National Amenity Societies, allowing them 28 days to comment;
   - works to be described in a public notice, allowing 28 days for public comment;
   - publicity in a local newspaper;
   - notificaton to the Royal Commission on Historical Monuments in England where demolition is proposed.
4. The decision-making body must take into account any representations arising from item 3 above and the desirability of preserving features of merit.
5. A fair and clear procedure must be established for settling disputes between the local congregation and the decision-making body.
6. Procedures must be established to deal with any breach of the internal control system, including provisions for reinstatements.
7. The process of internal control should be fully recorded in each case in which it is exercised.

8. Arrangements should be made to ensure proper maintenance, including a thorough regular inspection in a cycle of not more than five years.

Furthermore, the exemption was redefined as covering a church building (used primarily as a place of worship), any object or structure within a church building, and any object or structure either fixed to, or within the curtilage of, a church building and forming part of the land, provided that an object or structure of the latter two considerations is not listed itself, in the Ecclesiastical Exemption (Listed Buildings and Conservation Areas) Order 1994. In England the Order provides continued exemption, but on a reduced basis, for the Church of England, the Roman Catholic Church, the Methodist Church, the Baptist Union of Great Britain, and the United Reform Church. Details of these denominations arrangements and special arrangements in the case of Church of England cathedrals are to be circulated to LPAs in the course of time.

For denominations, faiths and independent congregations not defined in the Order the normal rules for LBC (and conservation area consent) have applied from 1 October 1994, although relevant requirements of worship authorities will be material considerations in the decision-making process. At the same time para. 8.7 of PPG 15 indicates that further Orders can be made if 'any other denominations or faiths are subsequently accepted as qualifying'.

With respect to proposals to fully demolish Church of England churches (which is not covered by the exemption), via the redundancy scheme of the Pastoral Measure 1983, the Church Commissioners have agreed to ask the Secretary of State whether he wishes to hold a non-statutory public inquiry where relevant bodies, including English Heritage, the Advisory Board for Redundant Churches, the LPA and the National Amenity Societies, have made an objection. Where it is determined that a building should not be demolished, the building will be vested in the Churches Conservation Trust (formerly the Redundant Churches Fund) to preserve it or efforts will be made to find a new use. The latter approach is strictly regulated by the relevant church bodies regarding whether a new use is appropriate.

Other powers to secure preservation

Apart from the relevant controls which regulate alterations and demolitions to listed buildings and protect their settings, other powers have been provided, as negative obligations, to assist in their preservation (Pickard, 1993c).

Criminal sanctions

Under s. 7 and s. 9(2) of the P(LBCA) Act 1990 it is an offence to carry out unauthorised works to a listed building or to fail to comply with a condition attached to a LBC respectively. The offence created by these provisons is one of strict liability and it is not necessary to prove *mens rea* (intention). This can be evidenced by the case of *R. v. Wells Street Metropolitan Stipendary Magistrates, ex parte Westminster City Council* [1986] JPL 903 in which the owner of a listed building authorised a contractor to remove 'everything of value', meaning only the furniture, but which was interpreted by the contractor as including fixtures and fittings. The contractor's defence that he did not know that the building was listed and therefore did not realise that he had committed an offence was held to be insufficient. What does have to be proved, however, is that the unauthorised works have affected the character of a listed building.

Defences to prosecution for the offence created by the above provisions may be found in limited circumstances under s. 8(2) and s. 9(3) respectively (see also para. 3.45 of PPG 15). It is also an offence to wilfully cause damage to a listed building under s. 59(1) of the P(LBCA) Act 1990. However, there is no defence for the offence created under this provision.

There are many case examples of successful prosecutions concerning unauthorised works to listed buildings. It is relevant to mention a number of the most significant publicised cases.

First, a developer was fined £1000 on each of 14 different counts for stripping out panelling and other interior features from a listed building in Dean Steet, Soho. By taking action on each unauthorised removal it was possible to go beyond the then maximum fine of £2000 available in the Magistrates' Court. Moreover, the defendant was also ordered to pay the costs of the LPA and it would appear that the fines could have been much higher but for the fact that the developer pleaded in mitigation that he had subsequently instituted an £80,000 repair scheme. In a second case brought before the Crown Court by Leeds City Council, a farmer was fined £7500 for demolishing a grade II listed barn and ordered to pay defence costs of £8500 (Bird, 1990). In a further Crown Court case the demolition of two listed cottages following the refusal of listed building consent resulted in a fine of £2000 plus costs of £1000 for the offending company and, more significantly, the two company directors were each fined £34,000 plus costs of £1100 (Cooling, 1990).

It would seem that the punishment meted out is designed to match the severity of the crime. Thus, in a case which received much publicity in the national press a developer was sentenced to four months in prison (and his accomplice 28 days) by Plymouth Crown Court for the bungled geligniting of a listed Wesleyan chapel. The plan had been to put a crack in the building, rendering it unsafe, so that there would be grounds to obtain permission for its demolition. However, the action misfired causing the front of the building to be blown out (Durham, 1992).

The relevant prosecution provisions, now found in s. 9(4) of the P(LBCA) Act 1990, were found to be insufficient to act as a deterrent against unauthorised works. The maximum fine on summary conviction was only £2000 and, while conviction on indictment could lead to imprisonment for a term of up to one year or an unlimited fine, there was a reluctance to seek trial by jury in the Crown Court because of the uncertainty of obtaining a prosecution. Although the evidence of the fore-mentioned case outcomes reveal that significant penalties can be made against offenders of the listed building regime, the government decided to strengthen the measures via the Planning and Compensation Act 1991.

In the Magistrates' Court the maximum fine was raised from £2000 to £20,000 and the maximum term of imprisonment was increased from three to six months. On conviction on indictment the maximum term of imprisonment was increased to two years. The maximum fine for contravening the terms of a LBEN was also increased from £2000 (and a daily fine of £200 for continuance) to £20,000. In addition, a new offence was created for the situation where a person obstructs a LPA from carrying out works required by a LBEN. The costs of such works can now be recovered from the owner through the County Court under s. 16 of the County Courts Act 1968. These changes were provided by Schedule 3 of the Planning and Compensation Act 1991 and came into being on 25 September 1991.

Failure to take the required steps indicated in a LBEN is an absolute offence (see *Mid-Devon District Council v. Avery* [1994] JPL 40). Moreover, while it is open to a person who is in breach of a LBEN to make an appeal under s. 39 of the P(LBCA) Act 1990, s. 64 of the same Act indicates that the validity of a LBEN may not be questioned in the courts.

Enforcement

Apart from taking criminal proceedings, LPAs may also issue a LBEN in respect of unauthorised works. Enforcement powers under s. 38 of the P(LBCA) Act 1990 are discretionary, allowing the LPA to consider whether it is expedient to take such action, bearing in mind the relative impact of any unauthorised works upon the character of a listed building. There is no time limit on the service of a LBEN.

Where works requiring LBC have been carried out without consent or where works have been carried out in contravention of a condition attached to a LBC, a LBEN may be issued specifying that the following steps be taken:

1. to restore the building to its former state; or
2. where such restoration would not be reasonably practicable, to execute other works, as necessary, to alleviate the effect of works carried out without LBC; or
3. to bring the listed building to the state in which it would have been if the terms and conditions of any LBC had been properly complied with.

The power to remedy a breach of listed building control is limited to the actual breach and cannot be used to alleviate previous unauthorised action not the subject of the notice in question, or require improvements to restore an historic appearance. For instance, in *Bath City Council v. Secretary of State for the Environment* [1983] JPL 737, enforcement action requiring that the complete roof of a listed building be recovered in Welsh slates, an improvement on the situation at the time the LBEN was served, failed. The roof had been patched from time to time with asbestos slate and corrugated slates. Thus the pre-existing condition had been one of disrepair in which restoration to the building's former state had not been reasonably practicable. However, re-roofing had eventually become inevitable and the roof was completely recovered in asbestos slates without LBC. The court found that the object of ground (2) above was to remedy the situation rather than punish the offender and therefore the enforcement procedure had not been appropriate in this case.

A LBEN must specify the date on which it is to take effect and a compliance period, and must be served on the owner (or other persons with an interest in the building who may be materially affected by it) not later than 28 days after its issue and not later than 28 days before the specified date. An amendment to s. 38 of the P(LBCA) Act 1990 created by s. 25 Schedule 3, para. 19 of the Planning and Compensation Act 1991 allows a LPA to specify different time limits within which the various requirements of a LBEN must be carried out. This provision also makes it possible to waive or relax the requirements of a LBEN without having to withdraw the whole notice, as was previously required. Penalties for non-compliance with a LBEN under s. 43 of the P(LBCA) Act 1990 are similar to those given for other offences under s. 7 and s. 9.

Injunctions

Unlike the provisions of the principal planning legislation (the Town and Country Planning Act 1990), there is no *Stop Notice* procedure in the listed building legislation. But where a LPA wishes to bring a swift halt to unauthorised works an injunction may be served. This power was extended to English Heritage and Urban Development Corporations, in addition to LPAs, via s. 44A of the P(LBCA) Act 1990. The evidence of a case example reveals the effectiveness of this power.

The case concerned the grade II listed Palace Mansion Hall, Newmarket. In March 1990 the owners (a property investment company) initiated work to the building including coating the exterior with adhesive and cement render without the required consent. The next day Forest Heath District Council obtained an injunction to stop the work. This was ignored by the owners. However, following the serving of a second injunction in April 1990 the director of the company was fined £25,000 for the offence. The council subsequently served a LBEN and a Repairs Notice under s. 48 of the P(LBCA) Act 1990 to restore the building (Suffolk Preservation

Society, 1990). (For further discussion of statutory repair powers, see Chapter 4.)

Rights of entry

The Secretary of State for the Environment may authorise rights of entry to relevant persons for the purpose of surveying listed buildings under s. 88 of the P(LBCA) Act 1990. In practice this power may be utilised by planning officials including conservation officers, as well as officials of English Heritage. The right of entry may be used for a wide range of matters but is perhaps most commonly used for inspecting buildings for the purpose of selecting new entries to the statutory list, reviewing the condition of a listed building for the purpose of seeking repair action, and to see if an offence has been committed against the legal protection regime. Under ss. 88A and 88B of the P(LBCA) Act 1990 it is a criminal offence to obstruct an official in the course of his duties under the rights of entry provision.

A case example in which English Heritage used its right of entry powers for the first time illustrates how this power can be used to assist the preservation of listed buildings. The case involved a pair of rare sixteenth-century smelting mills which had been in continuous use for some 300 years and had survived in a remarkably complete condition. However, they had fallen into a dangerous structural condition and, having failed to reach a mutually satisfactory solution with the owner of the mills, relevant English Heritage officials made entry to enable essential preservation works to be carried out.

The right of entry powers may be assisted by a general planning procedure provided by s. 1 of the Planning and Compensation Act 1991, whereby LPAs may serve a 'Planning Contravention Notice' for the purposes of obtaining information about activities on land where a breach of *planning* control is suspected. Representations by an owner, occupier or other recipient may be requested by a LPA. This power is also backed up by prosecution provisions as failure to comply with a notice within 21 days is a summary offence.

References

Bagnall, B. (1995) Advertisement regulations and listed buildings, *Context*, No. 45, March, p. 29.

Bird, C. (1990) Crime and punishment, *Conservation Bulletin*, Issue 10, February, p. 5.

Bold, J. (1992) Demolition revisited, *Context*, No. 36, December, p. 27.

Brereton, C. (1991) *The Repair of Historic Buildings: Advice on Principles and Methods*, English Heritage.

Calvocoressi, P. (1990) uPVC windows – a recent decision, *Context*, No. 28, December, p. 30.

Carter, H. *et al.* (1992) *Coming Unstuck: the Removal of Fixtures from Listed Buildings*, Victorian Society.

Catt, R. (1987) Façade retention: keeping up appearances in town, *Chartered Surveyor Weekly Refurbishment Supplement*, 29 October, pp. viii–ix.

Catt, R. (1993): Campaign for commonsense conservation, *Chartered Surveyor Weekly*, 28 October, pp. 22–3.

Cooling, P. (1990) Owning a listed building – an honour or a chore?, *Property Management*, Vol. 8, No. 4, p. 315.

Durham, M. (1992) Getting away with architectural murder, *The Independent on Sunday*, 12 April, p. 7.

Edwards, M. and Martin, J. (1995) Planning notes: learning by example, *Estates Gazette*, 18 February, pp. 122–3.

English Heritage (1990) *The Conversion of Historic Farm Buildings: An English Heritage Statement* (revised 1993).

English Heritage (1993) *Conservation Area Practice: English Heritage Guidance on the Management of Conservation Areas.*

Fidler, J. (1991a) Framing options, supplement to *Conservation Bulletin*, Issue 14, June.

Fidler, J. (1991b) Framing options update, *Conservation Bulletin*, Issue 15, October, pp. 13–14.

Field, M. (1993) Transforming a decaying farm into modern offices, *The Architects' Journal*, 24 March, pp. 25–6.

Henderson, A. (1989) The front line battle betwen past and present, *The Journal*, 25 October, p. 7.

Highfield, D. (1991) *The Construction of New Buildings Behind Historic Façades*, E. & F.N. Spon.

Millichap, D. (1993) Listed buildings and conservation areas – draft revised guidance, *Property Review*, October, pp. 398–401.

Millichap, D. (1994) PPG 15 – listed buildings and conservation areas, *Estates Gazette*, 5 November, pp. 229–31.

Morgan, A. (1991) Enlightenment in an old barn, *The Independent*, 21 September, p. 39.

Morley, J. (1903) *Life of Gladstone*, Vol. 2, p. 78.

Mynors, C. (1992) What is demolition?, *Context*, No. 35, September, p. 11.

Newcastle City Council *et al.* (1988) *The Newcastle Initiative: Grey Street Renaissance*, Newcastle City Council, The Newcastle Initiative and Tyne & Wear Development Corporation.

Pertee, C. (1991) Hazards in retention of façades and refurbishment – technical and structural considerations, *Architect & Surveyor*, February, pp. 10–12.

Pevsner, N. *et al.* (1992) *The Buildings of England: Northumberland*, 2nd edn, Penguin.

Pickard, R.D. (1993b) The need for listed building consent, *Property Management*, Vol. 11, No. 3, pp. 207–15.

Pickard, R.D. (1993c) Listed buildings: the strengthening of the powers of protection and prevention, *Property Management*, Vol. 11, No. 2, pp. 114–21.

RICS/English Heritage/IPD (1993) *The Investment Performance of Listed Buildings.*

Suddards, R. (1988) *Listed Buildings*, 2nd edn, Sweet & Maxwell.

Suffolk Preservation Society (1990) An incredibly sorry site, *Newsletter*, Winter 1990/91, pp. 1–2.

Chapter 3

APPLICATIONS AND APPEALS

Introduction

From the study of legal and planning controls affecting listed buildings it is appropriate to consider the practical and procedural aspects associated with LBC applications and appeals. Moreover, where an application for LBC involves partial redevelopment or change of use or any other work which falls under the statutory definition of 'development', such a proposal will additionally require planning permission. Furthermore, where major works schemes are proposed which may affect areas containing historic buildings, special procedures have been adopted for considering the impact of such proposals.

Applications for listed building consent

Pre-application discussions and advice

Since the strongest justification is required for the total or substantial demolition of *any* listed building, such a proposal will require early consideration of relevant evidence. Furthermore, due to the specific code of law designed for the preservation of listed buildings which require LBC for works affecting their character, it is a logical process to first attempt to determine that character and discover how it may be affected when making a proposal for change. This is an essential prerequisite to the making of a successful application for LBC.

It is not possible to define clearly the special interest for which a building has been listed. Yet the main areas which make up 'character' may be identified. All buildings have a form recognised in plan, section or elevation – a design concept; they have interiors as well as exteriors, are made up of a variety of materials, have detailing, a history and individuality. Most buildings have some form of commonality in being part of an era, area or type. Some may be unique or rare examples of particular building types. Some buildings have important social, historical connections or technological features associated with them. All these factors represent the main body of features, though not totally exclusive, which make up the character of a listed building. Any proposal which affects these factors would require LBC. The effect does not have to be detrimental, as even an enhancing effect would need LBC. Some changes may not affect the character, but in all cases it will be the LPA, in the first instance, that has to determine whether LBC is actually required.

The proposer of works to a listed building should seek advice on the likelihood of gaining LBC according to the relative merits of a proposal in terms of how the character will be affected. In order to avoid rejection at the first hurdle, advice should be sought from persons with expertise in this area, such as a building surveyor, architect or planning consultant with relevant experience of historic buildings who may be able to put forward a defensible solution. Pre-application discussions with the LPA conservation officer or other relevant official can avoid delay in progressing works and costs associated with aborted proposals. Moreover, para. 3.25 of PPG 15 specifically encourages owners of listed buildings to seek informal advice from LPA officials who may, in turn, direct them to other sources where they can get advice. English Heritage may also be willing to provide advice on individual buildings, especially where unusual problems are likely to be encountered. This may be particularly relevant with buildings of outstanding interest (grade I or grade II*) or where a proposal may have to rely on grant aid. The national amenity societies are generally willing to offer advice to individual owners whenever possible.

A number of specific areas of work where advice may be required, in terms of whether they will affect the character of a building, may be indicated:

Detailing

This is possibly the least appreciated area of listed building control, as may be evidenced by case decisions concerning window replacements. But detailing is important in the sense that the character of a historic building is represented by the sum of all architectural and historic matters of interest. As with window alterations, the replacement of cast iron gutters with uPVC equivalents, the capping off of chimney stacks or the removal of interior ceiling mouldings, for example, will all have an impact on the character of a building. Yet such alterations to detailng are usually totally unnecessary to the long-term survival of a building. Thus proper advice concerning such matters should lead the proposer of works away from so-called improvements when they would be detrimental to character.

Design of new work

The quality of design of new work is another element of detailing which requires prior consideration, particularly in the context of buildings of recognised refinement where failure to match quality, for example in relation to joinery and interior moulding work, may affect character. Discussions with relevant LPA officials on workmanship and the pedigree of contractor may assist to resolve misgivings about the effect of new work. Moreover, a conservation officer may be able to provide advice regarding a number of suitable architects or contractors with a good track record of work on historic buildings.

Repairs

The use of modern building construction techniques and developments in material technology may be viewed as improvements to a building. However, this will be disputed where such action affects the appearance of a listed building. Thus, while repair work does not generally require LBC, a repair which does not match the original may be regarded as an alteration. Advice is required in this context as unsympathetic 'repair' work may lead to enforcement action (see Chapter 5).

Re-use

With listed buildings that are in very poor condition, vacant and in need of comprehensive rehabilitation, it may often be viewed necessary to undertake complete stripping out of plasterwork and defective timbers, and to carry out floor strengthening, among other matters. However, the 'gutting' of features of architectural or historic interest will require LBC. Other avenues may need to be explored in order to maintain the special interest for which a building has been listed. This requires careful examination of the building to determine what features may be saved. For instance, modern practice in relation to the treatment of dry rot is to remove more timber than may actually be necessary to stem the condition. Moreover, resolving the access point of moisture penetration and leaving a building to dry out may enable a significant amount of the original timbers to be safeguarded. Specialist advice may have to be sought to resolve such problems, bearing in mind that applicants for LBC must be able to justify their proposals.

Seeking expert advice regarding works proposed to listed buildings undoubtedly eases the passage of applications for LBC through the decision-making processs. Moreover, would-be applicants have been directed to principles of conservative repair and philosophy which, if heeded, may lead to defensible proposals (see Chapter 5).

Categories of application

LBC is required for three types of work, namely demolition, alteration and extension. However, a different approach may be more useful to the applicant and LPA for the purpose of determining the time when an application is to be submitted. Furthermore, the extent and type of information which may be useful to support the application will vary according to the type of work proposed (Atkinson, 1989). In broad terms, applications for LBC fall into three main categories of work:

Proposals solely concerned with basic principles

In practice, this type of application will invariably be related to a proposal to remove a particular feature without any new detailed work being carried out, such as the removal of an extension to a listed building which in itself detracts from its character. A further example may be provided in the case of the proposed removal of a curtilage building which falls under the protection of a principal listed building, such as a derelict and redundant barn in the context of a farm complex which may be of debatable architectural or historic interest. Economic viability grounds, in terms of the cost of repair and/or the difficulty of finding a new use, may provide a plausible argument for demolition particularly where the location of the buildings is rurally remote. Moreover, there may be strong reasons for not allowing conversion to a residential or other use due to the detriment that may be created to the setting of the principal building.

All that is likely to be required to support the proposal in the first instance in this type of application are simple line drawings of the feature or building to be removed and its setting, a location plan and a condition report by a structural engineer, chartered surveyor or architect. Elevation drawings and photographs may also be useful to assist the LPA and any statutory or other consultees in making representations to the LPA. Proposals should be submitted at an early stage to establish whether the basic principle can be followed through to conclusion. Whether or not detailed costings for repair work or evidence of attempts to sell an ancillary building will actually be required will be dependent on the circumstances of each particular case and may be required at a later stage following consultations.

Proposals mainly concerning matters of detail

An example of this type of proposal may be indicated where a redundant building is to be found a new use. For instance, where a large textile mill building with regular window openings is to be converted into open plan office space, the principle of re-use will probably not require LBC whereas small-scale internal demolitions and alterations probably will.

The LPA is likely to require time to consider the effect of the proposed changes upon the internal character of the building, as well as the effect of possible requirements for car parking upon the building's setting. The applicant may best serve his or her interests by having early discussions, giving time for matters of detail to be negotiated with the LPA, prior to the submission of an application. Submission may be appropriate at the stage of Building Regulations application submission so that both sets of application can be considered at the same time.

The type of detailed work associated with this type of application is likely to include installation of new services, fire safety work (for which the Building Regulations 1991 allow flexibility in the case of historic buildings) (see Chapter 6), structural reinforcing, some internal partitions, and the replacement of doors and windows. The applications will need to be accompanied by appropriate drawings, to show the proposals in sufficient detail so that the LPA can consider how the changes will affect the character of the listed building, and a condition report.

Proposals involving principles and details

Cases where extensive work is proposed pose more of a problem to the applicant due to the fact that a balance must be achieved between supplying sufficient detailed information so that a decision can be reached and undertaking an extensive amount of work which may eventually become abortive if the ultimate decision is a refusal of LBC. Where major change is proposed the applicant may find it necessary to justify the works according to significant historic research in order to show respect to past alterations, sympathy in use of materials, that restorations are based on proven facts and that other principles of conservation philosophy are to be respected (see Chapters 5 and 6).

Unfortunately, the LBC procedures do not allow for outline consent as is allowed with applications for planning permission so that basic principles can be approved before detailed matters are clarified. However, under s. 17(2) of the P(LBCA) Act it is open to the LPA to grant LBC subject to a condition reserving specified details of works for subsequent approval. However, para. B.10 of Annex B to PPG 15 states that the initial application must give at least enough details to allow the LPA to assess the general impact of the proposals on the building before granting conditional LBC. A five-year time limit is imposed on any matters which are granted subject to a condition. (Examples of conditions acceptable in appropriate circumstances are further indicated in para. 3.22 and paras B.5 and B.9 of Annex B to PPG 15).

Accompanying drawings and plans

Under s. 10(2) of the P(LBCA) Act 1990 applications for LBC must be accompanied by sufficient particulars to identify the building to which it relates, including plans and drawings as are necessary to describe the works which are the subject of the application. In this context, three types of plans may have to be submitted by the applicant:

Survey drawings/plans of the existing building

Plans, elevations and possibly sections of the building, as existing, should accompany all applications for LBC. These should be accurate to avoid any delays in the decision-making process. Accuracy is required: to show the extent of an owner's title in the building and if it is linked to other historic buildings; to identify features in the building such as pannelling, fireplaces, cornices and door and window openings; and to indicate proportions such as in the glazing bars in a sash window. Accuracy may be obtained by preparing 'measured' drawings to normal architectural drawing conventions. The British Standard 1192, *Construction drawing practice* Parts 1, 2 and 3, and the RCHME's *Recording Historic Buildings: A Descriptive Specification* contain useful advice in this respect (Swallow *et al.*, 1993). The National Building Record, held by RCHME, should also be contacted to see if a completed survey of a building has been deposited, and if not, a completed survey can be deposited for future reference by others researching a building.

The provision of accurate measured drawings of the existing building is a preliminary to understanding the building which is important in terms of any proposals which may affect its character.

Overall proposal and layout drawings

Depending upon the size of the building, scales of 1:50, 1:100 or 1:200 are appropriate to represent the global view of the proposals. Existing and proposed plans may be presented in tandem to ease examination of the proposed changes by the decision makers. In this respect it is important to ensure that measurements are in the same form. Thus existing plans, if devised from archive sources, should reveal measurements in metric to match modern-day requirements for metric measurement as expressed in proposal plans.

Relevant notes should be provided to explain the type of materials or goods used and finishes to surfaces to be provided. Furthermore, notes should be provided to confirm the retention of particular architectural features and to indicate that any repairs to be carried out will match the existing position unless specifically stated. These notes assist the LPA in considering the attitude proposed regarding changes and provide a clear view to a contractor of what exactly is required.

Detailed drawings

Detailed work requires specific consideration as it may equally enhance or diminish the character of a listed building. Detailed drawings may be required in relation to new joinery, brickwork, stonework, leadwork or any other matters of detail. Joinery, in particular, should be shown in plan, section and elevation to indicate the relationship of proposed work to existing elements.

Photographic survey information

Photographs may also be submitted with drawings and plans to highlight building details. They may be particularly useful where partial demolition work is proposed by the applicant. Advice on how to present photographic evidence may be sought from offical sources (Buchanan, 1993).

Photographic means of measurement may also be used to provide relevant drawings of elevations. Rectified photography may be utilised to produce a true to scale print of a building's façade. Similarly, photogrammetry, a three-dimensional image produced by stereoscopic photography, is increasingly being used with computer-aided design (CAD) techniques to produce line drawings of elevations which are able to include variations in the recession of a façade (ICOMOS, 1990).

The handling of listed building consent applications

The various steps involved in the processing of LBC applications are indicated in the Town and Country Planning (Listed Buildings and Buildings in Conservation Areas) Regulations 1990 (the 1990 Regulations) and Annex B to PPG 15. More detail is provided in paras 81–88 of DoE Circular 8/87. Proposals to review procedures as indicated in the draft PPG 15 were not carried through to the final version as they would have required amendments to the Town and Country Planning General Development Order 1988 and the 1990 Regulations. It is intended that the direction procedures will be modified in the short term (see Appendix 7), however, at present the relevant procedures may be summarised as follows:

Certificate of ownership

Under s. 11 of the P(LBCA) Act 1990 the applicant must provide a certificate when the application is submitted, showing that the applicant is the owner of the property concerned or, otherwise, that the owner has been notified of the application.

The owner is defined as the freeholder or a leaseholder where not less than seven years of the lease remain unexpired. The form of the certificate is prescribed in Regulation 6 of the 1990 Regulations.

Initial advertisement and notification of application

Under Regulation 5 of the 1990 Regulations the LPA must advertise the application in a local newspaper and via a site notice displayed on or near the building concerned, and English Heritage must be notified of certain types of application (see below). The information supporting the application, including drawings, plans, photographs and other relevant documents, should be made available, with the application, for a period of 28 days, for public inspection. This enables amenity societies and members of the public to make representations against proposals. The RCHME and national amenities societies must also be notified of demolition applications with the same period of time being allowed for comment.

Consideration of the application

Any representations received regarding an application must be taken into account in the decision-making process. LPAs are required to make decisions as soon as possible after full statutory consultation periods have expired and normally within eight weeks of receiving a valid application, although in controversial cases extensions of time may be required. LPAs are free to refuse an application without reference to any other bodies.

Consideration to grant listed building consent

Where it is proposed to grant LBC different procedures must be followed depending on the location and the grading of a listed building, on whether works of demolition are proposed, and on whether English Heritage is required to have an input in the decision-making process.

Outside London the LPA must notify English Heritage and the DoE regional office of all applications affecting grade I and II* buildings and additionally notify the DoE regional office of relevant demolition proposals affecting grade II buildings (see Chapter 2). The Secretary of State for the Environment will then aim to decide within 28 days of notification, and after consulting English Heritage, whether to call-in the case for decision making.

Inside London LPAs must notify English Heritage about all applications affecting any grade of listed building. English Heritage must then notify the Secretary of State for the Environment within 28 days of all applications for which it is minded to direct the London Borough to grant LBC (as in the same cases indicated in relation to buildings outside London) or minded to authorise the London Borough to determine as it sees fit, so that the Secretary of State may consider calling-in the application. English Heritage may direct the London Borough to refuse the application and where it does so the London Borough may, within 28 days, notify the DoE to enable the Secretary of State to consider calling-in the application. The Secretary of State will normally aim to reach a decision within 28 days.

Issue of decision

The LPA has no jurisdiction if the Secretary of State has called-in the application for decision making. Otherwise the LPA is free to determine the application, subject, in London, to any direction given by English Heritage. However, through new arrangements English Heritage may have less involvement in the London Boroughs in the future (see below). The decision notice must then be issued in the prescribed form as indicated in Regulation 3(5) of the 1990 Regulations. If LBC is given for demolition, the RCHME must be afforded the opportunity to record the building.

The role of English Heritage

The main function of English Heritage regarding the application procedures for LBC is to advise the Secretary of State for the Environment on called-in applications and to provide advice or directions to LPAs in the circumstances indicated above. However, in 1993 the Secretaries of State for Environment and National Heritage agreed that English Heritage should devolve more responsibility to the relevant London Boroughs in decision making. Thus, through flexible authorisations, these authorities may be given power to determine applications concerning minor alterations and extensions to grade II buildings, as they are elsewhere in England.

All applications for LBC actually made by English Heritage, i.e. with respect to listed buildings in its ownership or guardianship, must be submitted to the DoE for decision making instead of being dealt with by the LPA.

Applications by local planning authorities

In the case of listed buildings in local authority ownership, applications for LBC must be submitted to themselves as LPA by virtue of Regulation 13 of the 1990 Regulations. Such applications are deemed to be called-in by the Secretary of State for the Environment for decision making. Applications made by county councils must be submitted to the relevant district LPA which must then send them to the DoE.

Late applications

Under s. 8(8) of the P(LBCA) Act 1990 an application for LBC may be sought even though works have already been completed. Such consent is unlikely to be granted merely to recognise a *fait accompli* and, even if LBC is granted, a prosecution may still be brought for the initial unauthorised works as the works only become authorised from the date of the consent.

Other procedures associated with listed building consent applications

Compensation for refusal or revocation of listed building consent

Under s. 27 of the P(LBCA) Act 1990 provisions were available for a person to claim compensation for the refusal of LBC or the granting of conditional LBC for proposed alteration or extension work. However, this provision was repealed by s. 31(3) of the Planning and Compensation Act 1991 as far as applications made on or after 16 November 1990 are concerned.

If it appears expedient to revoke or modify a LBC, compensation may be claimed under s. 28 of the P(LBCA) Act 1990.

Purchase notices

Where LBC has been refused, or granted subject to conditions, an owner may serve a Listed Building Purchase Notice on the LPA under s. 32 of the P(LBCA) Act 1990 requiring it to purchase the legal interest in the land if it can be established that the relevant decision has rendered the building as being 'incapable of reasonably beneficial use'. Further details regarding this procedure may be found in DoE Circular 13/83.

Development affecting listed buildings

Planning applications affecting listed buildings

Apart from the issue of development affecting the setting of a listed building (see Chapter 2), s. 66 of the P(LBCA) Act 1990 states that LPAs (and where relevant the Secretary of State for the Environment) are under a duty to have special regard to the desirability of preserving a listed building when considering whether to grant planning permission for development which may affect it. However, it is unlikely that this duty can be effectively carried out without the application for planning permission being considered at the same time as an application for LBC as, in the majority of cases, new development will affect the character of a listed building. Accordingly, para. 2.12 of PPG 15 advises that both applications should be submitted together. Where only a planning application is submitted, the equivalent information that would be found in a LBC application should accompany it and, in any event, the determination of a planning application cannot be taken as predetermining the outcome of a subsequent application for LBC.

Planning applications affecting listed buildings also bring a requirement to notify English Heritage. Outside London this applies to planning applications affecting grade I or II* buildings. Inside London English Heritage must be notified of planning applications affecting the fabric as well as the setting of all listed buildings.

As with planning applications concerning development affecting the setting of a listed building, applications actually affecting the building must be determined in the light of s. 54A of the Town and Country Planning Act 1990. While the need for LBC is not affected by this provision, the question of granting planning permission for new development, such as for change of use, must be considered in the light of any approved development plan policies. Accordingly, paras 2.2 and 2.8 of PPG 15 have advised LPAs of the need to have suitable policies in their plans which will give encouragement to the satisfactory re-use of neglected listed buildings and para. 2.18 indicates the need for development control policies to allow flexibility in dealing with applications for change of use. At the same time it must be emphasised that the statutory duty to have special regard to the desire to preserve, as indicated in the *Heatherington* case (see Chapter 2), must be of 'considerable weight' to the decision maker (Millichap, 1994).

Article 4 directions

Under Article 5 of the Town and Country Planning General Development Order 1988 (GDO) directions under Article 4 of the same order, to bring certain categories of permitted development under planning control, may be made by LPAs without the prior consent of the Secretary of State for the Environment if they relate solely to a listed building or to development within the curtilage of a listed building. There is one exception to this provision in the case of development to be carried out by a statutory undertaker (see Appendix 7).

Publicity and notification of planning applications

Under s. 67 of the P(LBCA) Act 1990, LPAs are required to publish details in a local newspaper and provide a site notice to advertise planning applications for development which, in their opinion, affects the setting of a listed building.

Major infrastructure projects and listed buildings

The Transport and Works Act 1992 and associated regulations made a significant change to the way in which certain major transport and other works schemes are authorised. These include proposals for the construction and operation of railways, tramways, trolley vehicle systems, other guided transport systems, inland waterways, and structures interfering with rights of navigation. The Secretary of State for Transport may make an order authorising schemes which would otherwise require a private Act of Parliament. Furthermore, applications for 'a scheme' may be assimilated with applications for LBC (as well as conservation area consent and scheduled monument consent) where these would be required to fulfil the objectives of a scheme.

The significance of this legislation may be indicated by reference to the example of the notorious 'clause 19' of the private King's Cross Crossrail Bill which caused uproar in conservation circles as, had it passed into legislation, it would have disabled English Heritage and national amenity societies from commenting on how the proposals affected matters of conservation interest in the area. Of greater consequence, 'clause 19' would have conferred power on the Bill's promoters to compulsorily acquire and demolish or alter any building in a defined area, irrespective of whether the building was listed. The effect would have been to circumnavigate the legislative code laid down to protect listed buildings. Fortunately 'clause 19' was removed from the Bill. The dangers of setting a precedent which was to allow blanket powers to demolish a whole swathe of urban townscape became all too clear (Scott, 1990).

Nevertheless, from time to time it may be expedient, for a scheme to be realised, to authorise the demolition of particular listed buildings. In such cases it is far better that the issues at hand are fully considered so that a balancing of interests can be achieved through due process. The 1992 Act and regulations and associated changes to heritage regulations have attempted to do this.

First, the Transport and Works (Applications and Objections Procedure) Rules 1992 (SI 1992/2902) specify publicity arrangements and procedures for the handling of objections where the Secretary of State decides not to hold a public inquiry under s. 11 of the Transport and Works Act 1992. Rule 3 provides for the notification of certain bodies having statutory responsibilities in particular areas of interest. Thus the national amenity societies have been given a role where heritage interests are to be affected. Similarly, English Heritage must receive notice of an intended application (and a copy of the subsequent application) where works proposed would affect a listed building (and a conservation area or a scheduled monument). It is also required that LPAs receive advanced notification, for instance, where 'the works' involve a change of use to a listed building, and must be sent a copy of the scheme application where it affects any part of their area.

Under the same rules the proposed scheme may, in certain circumstances, require an environmental statement in order to satisfy the requirements of EEC Council Directive 85/337 regarding its likely effects on the environment, and, in particular, its possible impact on the cultural heritage in terms of architectural, historic or archaeological matters.

Secondly, under the Transport and Works (Inquiry Procedure) Rules 1992 (SI 1992/2817) an inquiry or hearing, although only mandatory where there is an objection by the local authority or a person affected by a Compulsory Purchase Order (CPO), will almost certainly be held where there is a significant number of objectors to a scheme. Where objections are received in respect of an application for LBC (or other relevant heritage consents), the Secretary of State must consider these in deciding whether to hold an inquiry in the particular circumstances, though he is not obliged to do so.

The scheme applicant must also obtain any necessary consents, including planning permission where appropriate either separately or in parallel with the works order application. Under the Transport and Works Applications (Listed Buildings, Conservation Areas and Ancient Monuments Procedure) Regulations 1992 (SI 1992/3188) the Secretary of State may make provision for the assimilation of the procedures for making applications and holding inquiries. These regulations apply where application for relevant consents is made not later than 10 weeks after the works application or, if later, if the Secretary of State believes it to be appropriate. These regulations also modify the relevant regulations and legislations as regards the documentation which must be submitted with an application and the publishing of notices in local newspapers. Regulation 5 also allows provision for the holding of concurrent inquiries for both the works application and heritage consent applications.

Guidance issued by the Department of Transport further indicates that, in exceptional cases, there may be circumstances where it will be difficult to establish the precise effects of proposed works on listed buildings (and conservation areas), to grant a form of outline consent subject to a condition that the detailed proposals are submitted to the Secretary of State or LPA where relevant.

Despite these new provisions the Transport and Works Act 1992 does allow for a works order to be modified so as to exclude any statutory provision relating to any other matter (including heritage matters). But the relevant guidance suggests, by example, that the Secretary of State is likely to reject an application for modification in respect of provisions for the protection of the built heritage. Thus the new provisions should work towards greater cooperation between conservationists and promoters of works schemes so that the problems raised by 'clause 19' do not arise again (Collins, 1993).

In other respects the impact of new traffic routes and roads in centres or settlements have been given specific policy consideration in section 5 of PPG 15. Planning and highways authorities have been requested to set common objectives and consult each other about traffic proposals.

Appeals

Appeal provisions and regulations

Under s. 20(1) of the P(LBCA) Act 1990 an aggrieved applicant may appeal to the Secretary of State for the Environment within a period of six months against a decision by a LPA (a) to refuse LBC or a grant subject to conditions, (b) to refuse to approve details of works associated with a conditional

LBC, or (c) to refuse an application for the discharge or variation of conditions under s. 19 of the P(LBCA) Act 1990. In any appeal under this provision it is relevant to consider the provision contained in s. 16(2) of the P(LBCA) Act 1990 regarding the desirability of preserving the building or its setting. It is also possible to appeal on the ground that a building does not merit its listed status under s. 21(3) of the P(LBCA) Act 1990 and ought to be removed from the statutory list. However, this ground of appeal is not generally permitted when a building is subject to an appeal against refusal of LBC and also requires the involvement of the Secretary of State for National Heritage.

When planning permission has been refused concerning an application affecting the setting of a listed building or in respect of an application also requiring LBC, an appeal may be made under s. 78 of the Town and Country Planning Act 1990. This form of appeal must take into account the provisions of s. 54A with regard to approved development plan policies and the general duty contained in s. 66(1) of the P(LBCA) Act 1990 (to preserve the building and its setting). Furthermore, where a listed building lies within a conservation area, the duty contained in s. 72(1) of the P(LBCA) Act 1990, to pay special attention to the desirability of preserving or enhancing the character and appearance of the area, must also be considered. Where both planning permission and LBC have been refused, appeals against the two decisions will be considered at the same time.

It is possible to challenge an appeal decision, within a period of six weeks, in the High Court on a point of law. This may include an error in procedure that has caused substantial prejudice to a case, but the courts are not entitled to investigate planning merits as such.

There is also a right of appeal against a LBEN to the Secretary of State for the Environment under s. 39 of the P(LBCA) Act 1990 at any time before the notice comes into effect. There are eleven specific grounds of appeal against a LBEN. In brief terms these are that:

- the building does not merit listed status;
- the alleged matters have not occurred;
- the alleged matters are not a contravention of law;
- the works were necessary on health and safety grounds or were necessary to preserve the building;
- LBC ought to be granted for the works;
- copies of the LBEN were not served in the required manner;
- the requirements of the LBEN are excessive;
- the compliance period in the LBEN is too short;
- the works required would not restore the building to its former state;
- other specified works are excessive; and
- works to bring the building into a state where LBC would be granted are excessive.

The Secretary of State for the Environment has the power to correct any defect, error, or misdescription in the LBEN or vary its terms so long as no injustice is caused to either party. Under s. 65 of the P(LBCA) Act 1990 a decision relating to an appeal against a LBEN may be challenged on a point of law. However, any attempt to quash the decision in the High Court is only allowed by leave of the court. Furthermore, on such an application to the courts a LBEN may be ordered to remain in force pending the final outcome of the proceedings (but may be subject to the provision that the LPA must pay damages if an appeal is allowed). These provisions are designed to ensure that only legal challenges of merit will be taken to the courts.

Appeal procedures

LBC, LBEN, planning and planning enforcement appeals may be by written representations, public inquiry or, less usually, informal hearing. The procedure for the written representation form of appeal is governed by the Town and Country Planning (Appeals) (Written Representations Procedure) Regulations 1987. Informal hearings are governed by a non-statutory code of practice, as indicated in Annex 2 to DoE Circular 10/88. However, due to matters of national or local interest which are often raised in connection with listed buildings, it frequently results in an appeal being determined by public inquiry. Moreover, apart from minor cases a public inquiry serves to test evidence more satisfactorily. Inquiry procedures for appeals concerning LBC are governed by the same rules as planning appeals under s. 78 of the principal planning Act, namely the Town and Country Planning (Inquiries Procedure) Rules 1992. Inquiries may be conducted by the Secretary of State or a delegated Inspector (see also Schedule 3, of the P(LBCA) Act 1990). For LBEN appeals the procedures generally correspond to those in relation to ordinary enforcement notices and here the Town and Country Planning (Enforcement) (Inquiries Procedure) Rules 1992 apply.

The basic outline of the different form of appeal are indicated as follows:

Written representations

This is the simplest form of appeal and is usually quicker and cheaper than other methods. But the publicity and openness associated with an inquiry are not present.

DoE Circular 11/87 gives guidance on the operation of 1987 regulations concerning written representations. The format of submissions should include a description of the site and its location, a statement of relevant planning policies and additional statements required to clarify the reasons for refusal or grounds of appeal, and, in the case of applications for planning permission or LBC, all relevant documentation associated with the original application including representations made by third parties. Written statements made by the LPA (including an appeals questionnaire), the appellant and interested third parties are viewed by each party to allow for a reply by another party. Strict time limits for the submission of relevant information are imposed on the parties. Representatives of the parties may attend a site visit, the object of which is to clarify representations to the situation on the ground and not to hear evidence. The appeal decision may be made by taking into account only such written representations and supporting documents which have been submitted within the relevant time limits.

Informal hearings

As an alternative to the written representations procedure an appeal relating to minor issues may be heard via an informal hearing, particularly where there are little or no third party interests or no complex legal or policy issues to be considered. It is viewed principally as an alternative to the public inquiry method of appeal. Approximately 10 per cent of all planning appeals are conducted by this method (Stubbs, 1994).

Informal hearings are intended to save time and money while giving the parties a fair hearing. An inspector usually leads the discussion in an atmosphere which is intended to be more relaxed than a formal inquiry. There is no right to the cross-examination of witnesses but questions may be posed by participants throughout the proceedings. Statements from the parties are required 21 days before the hearing so that the inspector can review the arguments to be presented from a reading of the papers. It is not normally appropriate to read out the written statements and no new material should be introduced at the hearing. The appellant is normally requested to start the discussion, and may be invited to make any final comments before the hearing is closed. The hearing may be adjourned to the site so long as no party is put to a disadvantage, or the inspector may ask the parties if they wish to accompany him to visit the site after the hearing.

Public inquiry

The fundamental difference between a public inquiry and other forms of appeal is that it allows the testing of evidence. Anyone can write a statement, but the ability to defend or justify that statement under cross-examination is quite another matter.

The relevant procedural rules for enforcement appeals concerned with listed buildings differ considerably from planning or LBC appeals. The latter has a lengthy pre-inquiry procedure involving the submission and exchange of evidence of participating parties. It may also allow for a pre-inquiry meeting to deal with such matters as the identification of issues, presentation of plans and documents, the likely duration of the inquiry and the nature and exchange of evidence to be submitted. Such a meeting may be advantageous, as by agreeing facts the inquiry may be shortened and allow savings in costs.

The presentation of evidence at an inquiry is normally given by written proofs being presented by various witnesses. Natural justice is satisfied by the parties being given an adequate opportunity to comment on the evidence presented. Skilful cross-examination can be used to expose weaknesses in the opposing case and it may also be used to test the validity of facts or to narrow the issues in dispute. Expert witnesses may be called by the parties, who may also choose to be represented by an agent such as a barrister or other person experienced in advocacy. The inspector may intervene if the examination appears irrelevant or to protect witnesses, particularly where unrepresented or non-professional.

A site inspection will be made at some point in time. In appeals against refusals or conditional consents it is normal practice for an inspector to make an unaccompanied inspection of the site before or during the inquiry although an accompanied visit may also be made and may in fact be requested by the LPA, appellant or statutory party. In LBEN appeal cases, site visits are conducted on an accompanied basis. No representations may be made during a site visit but an inspector may ask questions of the parties that are entitled to be present.

Relevant information to be presented at appeal

Planning inspectors either make or influence the decision in the majority of appeals concerning listed buildings. It is therefore appropriate for the parties to direct their evidence to matters which are relevant to the inspector.

In this context there are certain relevant statutory requirements derived from the P(LBCA) Act 1990 which must be considered. These include the requirement to have special regard to the desirability of preserving the building or its setting with respect to applications for planning permission under s. 66(1) and LBC under s. 16(2), to consider the effect of unauthorised works on the character of the building in the case of enforcement action under s. 38(1) and the relevant grounds of appeal under s. 39(1); and where the appeal relates to a building located within a conservation area, special attention must be paid to the desirability of preserving or enhancing the character or appearance of the area under s. 72. Appeals concerning development proposals which have been refused on grounds that the proposal would have a detrimental effect upon the setting of a listed building must also be viewed against relevant policies set out in approved development plans under s. 54A of the Town and Country Planning Act 1990.

Furthermore, an inspector would expect to see evidence relating to the relevant general criteria and specific demolition criteria for granting LBC as indicated in section 3 of PPG 15. In addition to the *actual evidence* presented to support arguments concerning these criteria, further supporting documentation regarding the following matters may be relevant for the inspector to make a decision:

- *The importance of the building.* This may include the use of plans, elevations and sections and photographs of the existing building; early street plans; the list description; and relevant plan and policy documents.
- *The structural condition/state of repair.* This may include the use of dated photographic evidence of the decay, infestations and repairs already carried out; survey details; schedule of essential, emergency and long-term repairs required; and competitive tenders of a quantity surveyor's costings.
- *The description of proposals to alter, extend or demolish the building.* This may be supplemented by preliminary studies including alternative schemes considered and the original application drawings.
- *The effects of the proposals on the building itself, on its features or on its setting.* This may be highlighted by illustrations showing the appeal building and adjoining buildings following the proposed works as compared to the building in its present position.

Financial viability

Supplementary information will be particularly relevant in the case of demolition proposals. Appraisals should include total development costs set against net capital value of repair/refurbishment to continue the existing use, repair/conversion to a new use, and demolition/redevelopment; any offer of grant aid or correspondence relating to a grant aid application; and evidence concerning attempts to sell the building on the open market, including the sale price.

Replacement building

This may include details to enable a comparison between the design of the proposed replacement building with the existing building; development plan policies with particular reference to townscape issues and conservation area policies where the site is located in such an area.

Presentation of evidence

The same principles should apply to any party regarding the presentation of evidence at appeal, whatever the form of the appeal. Evidence should relate to clear facts and not hearsay, though an appeal is not a court of law and there is nothing to stop personal views being presented. Yet at inquiry there is a fundamental difference between the views of a witness and an expert witness called to provide information on particular issues in that the expert is permitted to express an opinion. The opinion must be the actual opinion of the witness and not the opinion of anyone else.

For the main parties teamwork is the key to success. Public inquiries are a specialist business which necessitate a team based on experience or recommendation. It is often appropriate to involve people who have worked together. For instance, it may be advantageous if an expert witness knows how his advocate will approach the case particularly at the stage of re-examination. It is also important to be able to understand the arguments put forward by other expert witnesses within the same team.

A well-structured proof of evidence reveals its strength. The proof should include everything that is relevant to the case and avoid repetition and unnecessary information. The various issues to be presented should be dealt with in appropriate chapters, divided into particular topics, and may include:

- qualifications and experience of the witness
- historical backgrond to the area where the appeal site is located
- the history, features, setting and importance of the building
- other listed buildings affected by the proposal (where relevant)
- conservation area affected by the proposal (where relevant)
- archaeological matters (where relevant)
- description of the proposal
- justification of the proposal
- impact of the proposal
- development plan policies
- government policy
- miscellaneous information
- conclusions.

Each chapter should be supported by other relevant documentary information presented in appendices (Mynors, 1994).

Participating parties at inquiry

Apart from the main participating parties to the appeal (the LPA and appellant, and where appropriate, their expert witnesses and advocate), English Heritage may also play a significant role in the appeal. This may be as an expert witness for the LPA or where an application is required to be notified to English Heritage. The national amenity societies may be represented in cases where they are statutory consultees although they rarely employ legal counsel due to the level of costs involved.

The fore-mentioned are the main players but other parties may appear at an inquiry according to the relevant procedural rules and the inspector may allow other persons to appear according to these rules.

References

Atkinson, T.G. (1989) Papers presented at a seminar organised by RICS Yorkshire Branch Building Surveyors Division and held at St Williams College, York, 11 May.

Buchanan, T. (1993) *Photographing Historic Buildings*, Royal Commission on Historical Monuments, H.M.S.O.

Collins, E. (1993) Conservation and the Transport and Works Act 1992, *Conservation Bulletin*, Issue 19, March, pp. 6–7.

ICOMOS UK (1990) *Manual on Research and Recording Historic Buildings.*

Millichap, D. (1994) PPG 15 – Listed buildings and conservation areas, *The Estates Gazette*, 5 November, pp. 229–31.

Mynors, C. (1994) Conservation evidence at planning appeals, *Context*, No. 41, March, pp. 26–7; No. 42, June, pp. 34–5.

Scott, M.G. (1990) The dangers of 'Clause 19', *Conservation Bulletin*, Issue 10, February, pp. 1–3.

Stubbs, M. (1994) Planning appeals by informal hearing: an appraisal of the views of consultants [1994] JPL 710.

Swallow, P., Watt, D. and Ashton, R. (1993) *Measurement and Recording of Historic Buildings*, Donhead, pp. 94–5.

Chapter 4

LISTED BUILDINGS AT RISK

Introduction

In the early 1980s concern grew at the number of legally protected listed buildings that were decaying through neglect. Many listed buildings were being demolished as they had no 'reasonable beneficial use' or stood vacant for want of repair or rehabilitation (Fidler, 1987). In 1980 Lord Montague produced a report on alternative uses for historic buildings which urged the need for policies for the nation's historic building stock (Montague, 1980). This revealed many successful conversions of listed buildings and gave a clear indication that no building need be demolished simply because its original purpose had become functionally obsolescent. When Lord Montague was appointed the first chairman of English Heritage (the Historic Buildings and Monuments Commission for England was set up in April 1984 following the National Heritage Act 1983), the matter of dealing with the problems of neglect was given a high priority (Ekersley, 1991).

The idea of forming a 'Buildings at Risk' Unit within English Heritage developed from action first initiated by the Civic Trust. In the early 1970s the Civic Trust submitted two reports to the DoE (namely *Financing the Preservation of Old Buildings* (1971) and *Forming a Building Preservation Trust* (1972)) which looked into the methods of financing the preservation of historic buildings and the potential role of building preservation trusts (BPTs) to deal with problem buildings. The origins of the Architectural Heritage Fund (AHF), which was set up in 1976 as a specific aim of the UK Campaign for European Heritage Year 1975, lay in the 1971 report. The AHF has since financially supported many BPTs in repairing and rehabilating historic buildings. Moreover, the Civic Trust's initiative led to the compilation of local 'risk registers' in various locations throughout the country. These enabled certain local authorities with limited grant-aid budgets to select buildings on which to concentrate their attention. However, since such action varied according to local priorities, further action was required to provide a consistent, widespread approach to dealing with buidings at risk and to coordinate local efforts in terms of national aid priorities (Fidler, 1987).

In 1986 English Heritage implemented a pilot survey, in cooperation with

Kirklees Historic Buildings Preservation Trust and funded by the Manpower Services Commission, to visit and assess some 3000 listed buildings situated in and around Huddersfield and within the confines of the Kirklees Metropolitan District Council administrative area. In fact, due to new buildings being added to the statutory list from the national resurvey of buildings, over 4000 buidings were assessed.

The area comprised a variety of listed building types, including mills and grand commercial buildings built from Victorian wealth that were under threat following the decline of the textile industry – the original source of that wealth. The survey logged the location of each building, the listing status, market status and ownership details, and assessed the nature of risk in terms of occupancy and condition of each building in the pilot area on a form devised by English Heritage. The information was logged in a computer record and an analysis produced to assist the local authority in determining what action should be taken regarding buildings determined to be 'at risk'.

The results from the pilot study indicated that 5 per cent of the listed buildings within the Kirklees area were at varying degrees of risk through poor condition, vacancy or, by their nature, unoccupiable. Moreover, the study confirmed the need for a national survey and the pilot scheme was refined for general use and launched in 1989 (Brand, 1992a).

Assessing the problem

The national survey

During 1990–91 a sample survey of 43,794 listed buildings was carried out in different parts of England by local authorities or other organisations such as BPTs. Buildings were assessed by external inspection to determine degrees of risk according to two factors, namely, condition and occupancy (see below).

English Heritage scale to measure the degree of risk

Condition	*Occupancy*	*Risk category*	
Very bad	Vacant	1/1A	Extreme risk
	Partially occupied	2	Grave risk
	Occupied	3/3A	At risk
Poor	Vacant	3/3A	At risk
	Partially occupied	3/3A	At risk
	Occupied	4/4A	Vulnerable
Fair	Vacant	4/4A	Vulnerable
	Partially occupied	4/4A	Vulnerable
	Occupied	5/5A	Not at risk
Good	Vacant	5/5A	Not at risk
	Partially occupied	6	Not at risk
	Occupied	6	Not at risk

Source: Buildings at Risk: A Sample Survey, English Heritage, 1992

The scale devised to measure the degree of risk indicates that buildings categorised 1–3 were found to be 'at risk' while buildings categorised 4 were vulnerable to being 'at risk' in the short term. 'A' categorisation denoted listed structures which cannot be occupied.

From the sample survey results, 7 per cent of England's listed buildings (approximately 37,000) were estimated to be at risk and nearly twice as many were considered to be in a 'vulnerable' position. It was also confirmed that the number of listed buildings at risk and patterns of risk varied between different types of building. There was found to be an association between the affluence of an area and the degree of risk with areas of the country considered to be less affluent having a greater proportion of buildings at risk. Rural areas, with a higher number of listed buildings than urban areas, were also found to have a greater percentage of buildings at risk. This confirmed the problem of redundancy of large industrial buildings such as mills where the industry associated with such buildings has declined and, similarly, in the context of the agricultural industry as represented by redundancy in farm buildings. Other building types having a high incidence of risk were found to be those used for power (wind and watermills), storage (warehousing) and ecclesiastical buildings other than Church of England buildings. The largest number of buildings found to be at risk per single building type were those originally in domestic use. Furthermore, a higher proportion of buildings at risk were listed as grade II or located outside of conservation areas (Brand, 1992a).

The sample survey covered approximately 10 per cent of all listed buildings in England. English Heritage indicated an intention to survey all listed buildings within England by 1995. This estimate has since been re-evaluated, but by April 1994 approximately 30 per cent of listed buildings had been surveyed with some areas of the country, such as Suffolk and Greater London, having been fully covered (Pegg, 1994).

Causes of neglect

There are many reasons why listed buildings fall into a state of neglect. However, it is not usually one factor alone which causes this, but a combination of factors in varying degrees depending on the particular circumstances of a building. The range of factors may be summarised as follows:

Repair costs

All old buildings are expensive to maintain and repair. The majority of listed buildings were constructed before modern building regulation standards were formulated and, resultingly, may have impractical arrangements for rain water disposal for example, or the design of areas may make access for maintenance difficult. With widely varying forms of construction it is more difficult to predict the full extent of repairs required. For instance, where structural weaknesses have been discovered in a building originally of timber-framed construction but which has been added to or strengthened by masonry work through the course of the building's history it may be difficult to predict the relative loads on the building and, therefore, how to correct the weakness. Moreover, once work has commenced for the purpose of remedying a certain defect further defects are often revealed which require more extensive work. For example, an apparently minor dry rot outbreak may cause extensive and expensive works if the spread of damage is found to extend to floor timbers holding in place an ornate plaster ceiling.

The listing protection means that a building owner will be required to use materials for repair which are appropriate to the building's character. Such materials may no longer be in normal manufacture and despite the recent growth in companies specialising in traditional material production the cost is invariably higher than modern mass-produced equivalents (Taylor, 1995). Architectural salvage may allow for the use of second-hand materials but even this presents certain problems regarding their condition, and high costs in cleaning, sorting and transporting from a number of sources. Apart from this there is a dilemma regarding the relative ethics of salvage in that some operators have been known to obtain materials by theft or unauthorised removal from existing historic buildings.

Furthermore, the level of craftsmanship required for the repair of listed buildings using traditional techniques adds to the cost of repair. In some areas there has been found to be a shortage of labour skilled in traditional work. While the Conference on Training in Architectural Conservation (COTAC) is actively supporting the development of courses in traditional work, at present the relevant skills and craft work are often only possessed by older and experienced workers whose work is necessarily meticulous compared to modern building techniques.

There are further costs relating to the level of professional supervision and advice regarding works to listed buildings. A high level of technical expertise may be required in relation to all but the most minor repairs to historic buildings and the supervision of works will require a greater degree of monitoring than with modern buildings.

Historic buildings will also require a greater degree of maintenance and running costs. Furthermore, additional expense may be created if a building is to be adapted via LBC for a new use in order to secure long-term preservation. This may include additional design costs associated with obtaining LBC, including special work for instance to satisfy the requirements of fire regulations in a manner which does not compromise the integrity of a listed building.

Moreover, research has shown that repair and maintenance costs for listed buildings may be up to five times higher than for unlisted buildings (Hudson, 1992).

Functional obsolescence and redundancy

The finding of the buildings at risk sample survey that many listed industrial and agricultural buildings were at risk is not surprising. The decline of the textile industry since the 1950s has led to the redundancy of many large Victorian mills, particularly in the Pennine belt (Binney *et al.*, 1979). Other industrial and storage buildings such as dockland warehouses have suffered the same fate due to financial and technological reasons. Similarly, changes in agricultural practice since the Second World War have resulted in redundancy in farm buildings with the reduction in the number of small-scale farms, changes in husbandry and relevant machine technology (Wade, 1990).

The abandonment of large country houses by the landed gentry and institutional users led to the formation of SAVE Britain's Heritage in 1975 as a campaigning body to 'save' many neglected and redundant buildings. Since 1975 SAVE has produced many reports highlighting the threat to different types of listed buildings due to redundancy, including: the Victorian hospitals and asylums with changes in health service practice; ecclesiastical buildings as the number of worshippers dwindle on a yearly basis; town-based office buildings as users require more modern and efficient premises and institutional investors fear the financial risk of investing in older buildings; old market buildings as shoppers prefer the ease of purchasing from modern superstores; theatres and cinemas due, initially, to a reduction in the number of attenders and more recently to the developments of video, multiplex cinema complexes and other modern facilities; and railway buildings from the financial streamlining of today's transportation services (Binney and Watson-Smyth, 1991).

Once buildings are regarded as being functionally obsolescent or redundant, repair and maintenance activity is halted and without an economic user it is difficult to prevent neglect. Moreover, while many structures can be easily and economically adapted, this option is frequently discarded.

Problems with financial aid and tax disincentives

In theory, every listed building is eligible for some form of grant aid. But with the number of these protected buildings now exceeding 440,000 there are limitations on the ability to give financial support for preservation work from public sources. The public purse is finite and the level of public funding will inevitably depend upon political considerations and the general health of the economy.

The principal public agency for grant aid, English Heritage, saw a reduction in its funding for repair grants between 1988 and 1990 of approximately 40 per cent (Hudson, 1992) but since 1990 the government gave a commitment to increase funding on a year-by-year basis (Page, 1991). Nevertheless, this finite resource has to be targeted each year as the level of funding can never really be guaranteed from one year to the next. Grant aid from local authorities is also indirectly from the public purse and the 'capping' of local authority budgets since the 1980s has effectively brought the amount of money available from this source down to a minimum.

The result is that grant aid has had to be prioritised. Thus a listed building is likely to be able to obtain a repair grant from English Heritage if it is classed as being of outstanding interest (grade I or II* buildings) under s. 3A of the Historic Building and Ancient Monuments Act 1953 (normally up to 40 per cent of eligible works), or if it lies within a 'Town Scheme' under s. 80 of the P(LBCA) Act 1990 (a mechanism set up in some historic centres and funded jointly with the relevant local authority and normally up to 40 per cent in total) or if it is prominent within a conservation area via s. 77 of the P(LBCA) Act 1990 (normally up to 25 per cent). In London special provisions allow English Heritage to assist in the repair of historic buildings via the Local Government Act 1985.

Thus, for the vast majority of listed buildings (those which are grade II) it is unlikely that grant aid will be available unless the building lies within a 'Town Scheme' or conservation area. Although there are other sources of finance which may be used in connection with listed buildings the amount of money is also limited by public finance constraints. These include the Sports Council for creating sports facilities (amount variable); the Ministry of Agriculture, Food and Fisheries for reinstating/repairing traditional farm buildings (up to 35 per cent); the Rural Development Commission for converting existing buildings to workshops or light industrial use (up to 50 per cent); and Housing Improvement Repair Grants from local authorities (up to 75 per cent). Other possible sources include government regeneration agencies such as: the English Partnerships Initiative and Urban Development Corporations; the National Heritage Memorial Fund; the English Tourist Board; the Museums and Galleries Commission; the Architectural Heritage Fund (which grant-aids feasibility studies but only provides loans for repair/rehabilitation work); and charitable trusts such as the Pilgrim Trust, the Monument Trust and the Getty Grant Programme (English Heritage, 1990). The European Regional Development Fund and European Social Fund provided by the European Union (EU) are also increasingly allowing scope for funds to be utilised for building conservation purposes in regions where development lags behind the EU average and where economic restructuring is required in older industrial regions (Bates, 1993).

Since the commencement of the National Lottery in November 1994 there is the first sign of significantly improved fortune for properties of heritage value. It has been estimated that the Heritage Lottery Fund, which is managed by the National Heritage Memorial Fund, will have about £1.8 billon over seven years to spend on heritage projects on top of all other existing official sources. Grants from the fund are restricted to public, charitable or non-profit bodies with an interest in heritage preservation or conservation such as BPTs. They are for capital projects usually costing more than £10,000 where the applicant can raise a significant element of partnership funding in relation to the purchase, improvement or restoration of an existing 'important' heritage asset (MacDonald, 1994).

However, apart from the current shortage of finance (leaving aside the potential of the Heritage Lottery Fund) there may be other problems associated with grant aid. For instance, while a certain percentage may be made available to assist repair work, the majority of costs have to be met by the grant applicant. Furthermore, the work will have to be approved, and in the case of conservation grants this may mean strict control over how the work is carried out – particularly in relation to materials used, workmanship and professional supervision. Resultingly, the grant may not be paid until the work has been inspected and approved. The applicant may be further constrained by the need to obtain competitive tenders which have to be submitted to the sponsor for scrutiny before an award is agreed, or risk delay or uncertainty regarding the level of grant given. Added to this is the possibility of an agreement as a pre-condition to funding to allow public access once the works are completed (Davey, 1992).

The onus is on the applicant to provide all necessary drawings, specifications, schedules of work and so forth. Applications for grant aid, therefore, are not straightforward. Moreover, the web of powers, regulations and conditions associated with grant applications frequently makes the task formidable even for professional advisers and can actually deter owners of listed buildings from pursuing repair work.

A further problem may be indicated in relation to the Value Added Tax (VAT) incentives for listed buildings as detailed in H.M. Customs and Excise VAT Leaflet 708/1/90 (see Chapter 6), which actually works against the proper repair and maintenance of protected buildings. This is due to the fact that when general building work was included as a cost upon which VAT would be levied by the government in 1984, approved alterations to listed buildings were zero rated whilst all other works of maintenance and repair were not. Owners of listed buildings have been discouraged from undertaking conservation works by minimal intervention. Moreover, the 'incentive' indirectly gives encouragement to the neglect of building so that owners may argue the need for drastic amendments of existing features or even major reconstruction works such as a façadism proposal as a means of gaining a tax advantage (Nelson, 1991).

This anomaly has undoubtedly worked contrary to the 'desire to preserve' found in official policy. Moreover, it has been the subject of critical discussion in conservation circles for many years. Yet the only opportunity to change the current system of VAT relief would appear to be when certain European tax legislation is harmonised, as envisaged by the Single European Act in 1986. This was first proposed to take place by the creation of the single European market on 1 January 1993 but logistical problems have resulted in a transition period until what is known as the *Origin System* can be implemented. The target date for the harmonisation of VAT rates proposed within two bands under the *Origin System* is 1997. It is envisaged that this will allow for the majority of goods and services to be taxed at a standard rate with the possibility of a lower rate for certain agreed items. This would then provide the opportunity for creating VAT relief for the maintenance of protected buildings of architectural and historic interest under a broad definition accepted by European Union member states. The need for such approaches was recognised in the European Commission's *Green Paper on the Urban Environment* published in 1990. Paragraph 5.3 indicated the need to provide more substantial financing for the conservation of historical buildings and areas of European significance. Further examination of these issues is proposed via the European Union's *Raphael* programme from 1996 (Burman, P., Pickard, R.D. and Taylor, S., 1995).

Value and tenure

There is frequently a disparity between the cost of repair of listed buildings and the end value of listed buildings once repaired. Even with the possibility of grant aid, a conservation project may still be perceived to be, or actually be, uneconomic.

In some cases the price paid for a building, if bought speculatively, may not fully consider the level of repairs involved or how significantly the legislative requirements affecting listed buildings are in terms of controls over works as compared to unprotected buildings. While PPG 15 has indicated a much stricter marketing test for LBC applications for demolition so that buildings must be offered for sale at a price which reflects the condition of the building, this may encourage owners to sell the property for a higher price rather than seek such consent or undertake the necessary works.

Even if a building is repaired or rehabilitated it may only be attractive to certain types of users who are prepared to accept non-standard accommodation. Moreover, research carried out in 1993 on the investment performance of listed buildings highlighted the fact that institutional investors in property have traditionally only been willing to invest a small percentage of their funds in listed buildings despite the fact that this research has dispelled the myth that listed buildings never perform as well as modern equivalents in investment terms (RICS/English Heritage/IPD, 1993).

There has also been the problem created by the auction of property where unwanted buildings are included within a package of property folios for sale. Such property may be found in many cities in England. Examples found in the Grainger Town area of Newcastle upon Tyne include nineteenth-century listed buildings now in secondary positions with vacancy in upper floors. With landlords being absent from the area and no organised maintenance or management policy, the under-performance of such buildings in investment terms eventually lead to these buildings becoming 'at risk'.

Modern letting arrangements may also have a detrimental impact upon listed buildings with common institutional practice being to let commercial property on fully repairing and insuring leases for up to 25 years. The tenant's responsibility to repair may not be carried out until the end of the lease when a schedule of dilapidations will be assessed for the tenant to pay in lieu of a regular maintenance programme. Not only does this encourage the neglect of a building but there is no guarantee that the money eventually received will be used to repair the building. Moreover, where the amount of costs is disputed, the building may be left to decay further until negotiations or court action can settle the amount of money that is required to cover the cost of repairs.

The provisions of the Leasehold Reform Act 1967, which were extended by the Leasehold Reform, Housing and Urban Development Act 1993, allow certain qualifying tenants of houses and flats to have a statutory right to enfranchise their interests. The effect of these provisions may deter some owners from letting their property, causing some buildings to remain empty. More significantly the 1993 Act widens the scope to enfranchise, which may cause the break up of some of the grand residential estates of houses in London that have been collectively managed, allowing individual owners to make alterations such as the repainting of external façades in different colours previously prevented by lease covenants.

Planning and other regulation matters

Since the priority for development control decisions has been placed firmly in favour of development plan policies under s. 54A of the Town and Country Planning Act 1990, it is now open (and is actively encouraged by government policy) for LPAs to establish planning policies which safeguard conservation interests. Yet until this change in government policy in 1991 this link was not always established. In some instances restrictions have been placed on the re-use of listed buildings or re-use has been prohibited by strict car parking standards. The strictness of green belt policies have also caused disputes regarding the re-use of redundant listed buildings, and with the consequent delay some buildings are left to decay.

New road schemes and other planning decisions have also blighted areas, and these have had a consequent effect upon listed buildings in such areas.

Access problems may limit the potential of re-using redundant listed buildings particularly where they are situated in remote rural locations.

Building regulations, particulaly fire safety regulations, may limit the potential of re-use unless a flexible attitude to such matters can be negotiated between owners, relevant planning or conservation officials and building inspectors (Pickard, 1993d).

Insurance

There is no obligation to insure a listed building. Insurance costs for listed buildings, particularly large ones, can be very high. This is due to the fact that with a listed building an owner may be required to reinstate or repair the building beyond the point that would normally be considered economic or reasonable. This could have two possible impacts. First, an owner does not insure the building at all, thus resulting in abandonment in the event of a major fire. Secondly, that minimal cover is provided on a first loss basis which still leaves the building at risk if a major damage is incurred (see Chapter 6).

Area deprivation

Many towns and cities have suffered from the decline of traditional industries. Redundancy and obsolescence should not be viewed as isolated issues in this context. Historic buildings may become at risk throughout an area due to the loss of business activity and employment. This has a consequent effect on all property sectors. Moreover, investment initiatives are often directed to new development on the premise that a new face is needed to bring life back to an area and the perceived view that old buildings cannot provide the functional space required for modern business activity.

The reluctant owner

Lastly, an owner of a listed building may just not wish to keep a building in repair for whatever reason. Many owners of listed buildings still hold to the view that 'an Englishman's home is his castle', despite the fact that such buildings are now protected by legislation. No one can force an owner to carry out repairs even if there are powers available to persuade an intransigent owner.

More significantly, owners may be reluctant to take the required action to prevent a building becoming 'at risk' due to the interaction of some of the above factors.

Dealing with the problem

The results of English Heritage's sample survey confirmed that in order to reverse the decline of historic built environments and resolve the particular problems regarding unused and decaying buildings, agreements would have to be reached between owners of listed buildings, local authorities, English Heritage and amenity societies about the best strategies to employ (Brand, 1992b). To some extent the English Heritage initiative paved the way for a national focus upon area conservation strategies, the idea for which had commenced from Civic Trust initiatives in the 1970s and 1980s. Moreover, these examples and the buildings at risk campaign in turn paved the way for local authority initiatives for individual buildings and area conservation–regeneration programmes, and a new form of partnership approach to reversing the fortunes of the deteriorating historic building fabric on a much wider scale.

The development of the area strategy approach

The small town of Wirksworth, once the third largest town in Derbyshire, can be seen as the birthplace of the area conservation–regeneration movement. In 1978 the Civic Trust launched a project in response to a question raised by the Monument Trust as to whether anything significant could be achieved if the Civic Trust focused its funds in a declining area initially for a period of three years (Michell, 1989a). The town had a few surviving medieval buildings, eighteenth-century merchant and town houses, and historic inns and public buildings of recognised quality built from wealth created from agriculture, mining and quarrying. But many of these were neglected in the mid-nineteenth century following the opening of a quarry near to the town and other construction works which eventually throttled the town. Despite continuing decline during this century, the decaying built fabric remained largely intact.

The Wirksworth Project commenced with the employment of a project officer with the idea of involving the local community in revitalising the town's attractive, if neglected, historic building fabric. This essentially was to be a loosely created partnership between local people, local, county and central government, the Civic Trust and other bodies with conservation and economic credentials. Prior to the launch of the project the local authority set up a conservation area in part of the town and a 'general improvement area' as a mechanism to improve the decaying housing stock. Once the project commenced it extended the conservation area to include the slopes of hills surrounding the town, pledged one-sixth of its annual budget to the project, created a community information centre, and with central government set up a 'Town Scheme' with a budget of £30,000 p.a. towards conservation projects. The project officer coordinated other activities including raising community interest in the project through a newsletter and other publications, and encouraging the local civic society to develop a town trail and practical environmental improvements, stimulate investment in shops, encourage employment, and develop the town's visitor potential (Pearce, 1989).

The Civic Trust's direct involvement in Wirksworth came to an end after three and a half years but by then the momentum of the project was such that the local authority decided to continue the initiative and it did so with considerable success. This was achieved through encouraging by example and partnership. The Abbey National Building Society agreed to fund a demonstration home to show how improvements could be carried out in a sensitive manner and gave preference to mortgage applications in the conservation area. The Derbyshire Historic Buildings Trust (DHBT), one of the county's most successful rescuers of 'hopeless cases', took responsibility for four important building repair projects. In one scheme involving an important seventeenth-century house in a ruinous state within the centre of the town a sensitive reconstruction was carried out with the aid of documented evidence and with the help of grants from the Development Commission and Derbyshire County Council to meet the £185,000 total cost (Figs 4.1 and 4.2). DBHT's commitment to the town was such that they decided to take over the project's priority list of 16 important buildings at risk, and, through the encouragement of the Monument Trust, to set up a revolving fund to specifically deal with these buildings which, in turn, enabled them to establish a full-time presence in Wirksworth.

Other partnership and exemplar activities included the conversion of a derelict building into a heritage centre; new economic investment in office factory accommodation by the County Council, the Development Commission and COSIRA; the conversion of a redundant clinic to house new offices for COSIRA; investment in retailing through the provision of new shops and upgrading existing shops with traditional shop front design; the repair of many buildings and traditional street surfaces through grant aid; and the conversion or re-use of redundant buildings into an art gallery, a bistro, workshops, offices, shops, housing and a chemical business (Fig 4.3). In 1983 the project was given a Europa Nostra award.

The Wirksworth Project initiative was cost effective in bringing considerable investment from other sources. One aspect of this was the grant aid to buildings at risk provided through the 'Town Scheme', which in turn generated a significant amount of investment in listed buildings. For example, the 'Town Scheme' budget of £24,000 p.a. for the years 1986 to 1988 led to investment costing approximately £160,000 for each of these three years (Michell, 1989b). In 1993 English Heritage reviewed its financial commitment in the 'Town Scheme' and decided that the highly successful and pioneering scheme had fulfilled its intentions and should be terminated (Johnson, 1993a).

The success of the Wirksworth area strategy approach led the Civic Trust to develop further small town initiatives. However, a significant broadening of the idea was developed in 1984 to reverse the declining fortunes of a much larger industrial town. This was the textile town of Halifax and the surrounding Calderdale area which had a decaying architectural legacy of mills, warehouses, shops, houses and Victorian municipal buildings.

The area's renaissance began in the early 1970s when the local council saved from demolition, by one vote, the Georgian colonaded Piece Hall. Nearly £300,000 was spent on rejuvenating the 200-year-old grade I listed building by creating 40 shops, an art gallery, a museum, a tourist centre and an open market (Fig. 4.4). Calderdale M.B.C. also made a commitment to restoring the past glory of the town by implementing a stone-cleaning pro-

Fig. 4.1 1–3, Green Hill, *c.* 1631. Once one of the most important town houses in Wirksworth, in 1954 the roofs of numbers 1 and 2 fell in, causing the lower floors to collapse with them, but leaving the walls remarkably intact. (*RCHME Crown Copyright.*)

Fig. 4.2 1–3 Green Hill, Wirksworth. Repaired and reconstructed 1981–3 on the basis of research evidence, and rehabilitated to provide office space by the Derbyshire Historic Buildings Trust at a cost of £185,000.

Fig. 4.3 Art Gallery and other historic buildings in the Market Place, Wirksworth. This follows repair works in 1980 and later streetscape improvements which further enhanced the setting of many buildings individually improved by their owners.

Fig. 4.4 Section of the preserved eighteenth-century grade I listed colonaded Piece Hall, Halifax; now in a variety of uses.

gramme on a massive scale, stripping away more than a century's industrial grime from historic buildings. But in 1982 the grade II listed Dean Clough textile mills, the largest mill complex in Europe, closed with the loss of over 600 jobs. The textile industry had been in decline since the 1950s, but local people reacted as if this was the death knell for the area.

In 1984 the Civic Trust produced a report entitled *Halifax in Calderdale – A*

Strategy for Prosperity which led to the development of the Calderdale Inheritance Project (CIP). The idea behind this project was to create a 'community partnership' for the 'holistic regeneration' of the area. As well as attracting new industry and new development, the project aimed to use the man-made historic environment as an asset and basis for revitalising the area (Carvill, 1984). This was to be implemented via three interlinking campaigns (CIP, 1987):

Partnership

The purpose of this was to reduce bureaucracy and to create a positive climate for the public and private sector to work together. By example, one of the successful partnership schemes which was implemented was to regenerate the derelict town centre site known as Upper George Yard, which had a number of empty old buildings of architectural quality surrounding it. In normal circumstances – land assembly either by the use of compulsory purchase powers or piecemeal acquisition of sites – it was considered that it would have taken at least eight years to assemble and regenerate the area. However, the CIP and Calderdale M.B.C. pump-primed the project with an investment of £40,000 and persuaded the brewers Joshua Tetly to donate £1 million and some of the derelict buildings. A public house was restored with the addition of a new restaurant extension and a five-storey redundant industrial listed building was provided with an English Heritage grant to assist in its conversion to retail, restaurant and office use. Environmental ground improvements were made by the council in installing paving materials which were donated by local firms. Moreover, the interest created by these works led adjacent property owners to make repair improvements to their premises in order to take advantage of the commercial opportunities created by the scheme. Thus, by efffective cooperation the area was regenerated, respecting the historic qualities of the area, in approximately four years.

Exemplar projects

The purpose of this was to create an impact through projects in key locations as a catalyst to encourage building improvements. One such example was a scheme to encourage the restoration of traditional shop front designs by an exemplar scheme associated with the Victorian borough market shops (Fig. 4.5), the production of a shop front design guide which identified the design of lettering styles for Victorian shops and the provision of grant aid for listed retail premises in this context. Planning officials and CIP team members assisted applicants in developing designs that were more in keeping with the architecture of existing buildings.

The response to this initiative was significant, even extending to retail premises designed after the Victorian period. For instance, when McDonalds acquired a former clothing shop in Art Deco style they were persuaded to change their normal corporate design finish for a new shop front in Art Deco style.

Fig. 4.5 Borough Market shop fronts, Halifax. An exemplary scheme to encourage the return of traditional shop front designs.

Further exemplars were provided by private initiatives in keeping with the aims of the CIP. The most significant of these was the visionary scheme of an entrepreneur who purchased the listed but redundant Dean Clough mill complex for new business space, including workshop, printing, office, warehousing and distribution space, an 'enterprise campus' and a gallery of contemporary art (Fig. 4.6). The complex now has a thriving community of musicians, sculptors, printmakers and painters, thus blending art with business in the kind of utopian ideal of which the nineteenth-century designer and conservationist William Morris would have approved (de la Hay, 1993). The obvious success of this project in turn led to further schemes to re-use other derelict mills in the Calderdale area.

Promotion

Promotion of the regeneration project was seen as necessary to attract grant aid, other sponsors, supporters and clients. The CIP and the council carried out a variety of public relations exercises through marketing, publications, workshops and exhibitions. The work of the CIP attracted interest from H.R.H. the Prince of Wales who as President of 'Business in the Community' introduced a high profile to the 'Calderdale Partnership'. Business forums to attract investment and boost confidence were also arranged.

In September 1988 Calderdale M.B.C. was host to a unique Council of Europe conference on successful regeneration attracting delegates from over 20 countries. The fortunes of the area had transformed in a significant way. The re-use and repair of the architectural fabric of historic properties of the area had been the main ingredient in promoting the area as an excellent location for living, shopping and working. Moreover, this was reflected in a rise in property values with, for example, office space in Dean Clough attracting rents on a par with or even greater than larger towns elsewhere in northern England.

Other regeneration projects

Both the Wirksworth and Calderdale projects were influential in spawning other areas and community-based conservation–regeneration projects elsewhere in the country with the assistance of the Civic Trust. Furthermore, in some cities the idea of using redundant or run-down listed buildings to regenerate particular areas have had a measure of success. Examples may be given in the Victorian Quarter, Leeds, including the restoration of traditional design schemes in Victorian arcades and the Kirkgate Market and the re-use of the grade I listed former Corn Exchange as a speciality shopping centre (Figs 4.7 and 4.8); the Lace Market, Nottingham, involving the re-use of nineteenth-century merchants warehouses and other listed buildings for office, residential, retail and textile workshop space; the Jewellery Quarter,

Fig. 4.6 Dean Clough. Once a centre of carpet manufacturing, this grade II listed former mill complex now houses a variety of uses.

Fig. 4.7 Exterior of the grade I listed Corn Exchange, Leeds, designed by Cuthbert Brodick in 1863.

Birmingham, an area dominated by nineteenth-century jewellers' workshops and other listed buildings where a mixture of uses have been introduced including recording studios, restaurants, housing and design studios; and Little Germany, Bradford, involving nineteenth-century textile mills re-used for office, exhibition, workshop and retail space.

Moreover, the development of these strategies has been assisted by support from English Heritage in redirecting grant funding away from traditional historic centres such as York and Chester to depressed industrialised centres. For example, at the commencement of the CIP in Calderdale, English Heritage provided a grant budget of £80,000 p.a., but once the

Fig. 4.8 Interior of the Corn Exchange, Leeds, following a £6 million investment to sensitively convert the building into a speciality shopping centre in 1990.

regeneration initiative was developed, funding was raised to £200,000 p.a. Positive encouragement for the development of area conservation strategies was given in an English Heritage discussion paper in 1987 at the same time as the 'Buildings at Risk' strategy was being developed (Robshaw, 1987). In turn, some city LPAs started to develop major conservation strategies based upon conservation areas and listed buildings at risk.

However, some LPAs have used their statutory powers to designate a greater number of conservation areas in order to attract funding from English Heritage with only a small percentage of 'outstanding' listed buildings otherwise attracting grant aid for repair work. For instance, by 1989 the city of Bradford had designated over 50 conservation areas whereas the city of Newcastle upon Tyne had only 7. While some would argue that this has denigrated the idea of conservation area policy, the greater demand for limited funding by this action has led to the need for more sophisticated strategy initiatives and a reassessment of how best to target grant aid by English Heritage. In line with this, the city of Newcastle upon Tyne developed a conservation strategy in 1989 which first aimed at dealing with the specific problem of buildings at risk before widening the scope for conservation area designation. Moreover, English Heritage supported this action by making Newcastle one of 14 pilot areas for a new type of 'conservation partnership'.

Building on the 'at risk' campaign

By undertaking surveys of listed buildings within local authority administrative areas, specific action may be devised for particular buildings found to be most at risk which otherwise may be untouched by a more general regeneration strategy.

In London, English Heritage established a *Register of Buildings at Risk in Greater London* which identified that 950 buildings were at risk by 1991 and work has continued to update the register since then. Each London Borough Council has been circulated with detailed guidance on appropriate ways and means of tackling the problem. This has included bringing pressure upon recalcitrant owners to repair or sell their properties using statutory powers where necessary. Furthermore, the need to forge closer links with BPTs and other public, private and voluntary organisations to coordinate plans for the promotion and marketing of vacant buildings and sensitive adaptions for new uses has been advocated. Moreover, English Heritage has sought a closer dialogue with large property-owning bodies such as health authorities, who own many large Victorian hospitals and asylums which are no longer regarded as viable to present health service requirements, so that pre-emptive measures can be made to prevent buildings falling into a state of neglect when the question of redundancy arises (Davies, 1991).

Many LPAs outside London which conducted a 'buildings at risk' survey devised their own methods of tackling the problem with assistance from English Heritage. By example, the Kirklees pilot study identified 197 listed buildings, or 4.4 per cent out of a total of 4470 surveyed, to be at risk. However, the majority of buildings at risk were listed grade II and only 40 per cent of these were located within a conservation area. Thus, under the prevailing conservation funding schemes the hard core of buildings found to be at risk were those least likely to receive any grant aid. It was not considered to be practicable or desirable to enforce preservation action under the repair provisions of ss. 47, 48 and 54 of the P(LBCA) Act 1990. Instead, an effective response was made in 1988 by the local authority setting aside a sum of £50,000 p.a. as 'historic building grants' which, in the first year, was targeted with £10,000 being allocated to each of five of the grade II buildings found to be at extreme risk. This money was ingeniously administered via the Kirklees BPT which worked to prevent the reclamation funds not employed during the year and allowed these moneys to increase compoundly until allocation.

The strategy employed with these buildings was to contact the owner to advise that the council was concerned about the condition of the buildings, identify the action required to safeguard further deterioration, inform the owners about the possibility of financial assistance for repairs or rehabilitation or offer help in finding new owners if the buildings were surplus to requirements. The reaction to this notification determined the subsequent course of action with the threat of imposing statutory repair powers if a negative response resulted.

The desired response to this action was to bring new life to these buildings which could act as exemplars to other owners. However, a review of the progress on the first five buildings after five years revealed a rather varied response:

1. A semi-redundant railway station became subject to an unconventional scheme to re-erect it following its acquisition for a nominal sum by two council-located purchasers as part of a proposed light railway scheme, British Rail having been unwilling to coordinate a scheme of repair. While the scheme is likely to be successful following the granting of a light railway order, the delay caused the buildings to further deteriorate and in 1991 one of the buildings was destroyed by fire.
2. A former warehouse, gutted by fire in 1986, attracted the interest of a property developer through a periodical publication concerning its plight. The potential to rehabilitate part of the remaining buildings was realised through the assistance of the £10,000 historic buildings grant and a further grant of £20,000 provided by the council's Economic Development Unit which led to a successful conversion to office premises. While the remaining part remained in a derelict state, the rehabilitated part provided a model exemplar (Fig. 4.9).
3. The owner of an abandoned farm was subsequently traced and coerced into drawing up a rehabilitation scheme. However, when progress on the project was not sustained a Repairs Notice was served on the owner, the outcome of which was that the owner decided to sell the property. The new owner commenced a partial restoration and re-use scheme without the targeted grant aid. Much of the work was reconstruction, some of which gained LBC but other work had not been approved. This then left a dilemma for the LPA as to whether they should take enforcement action.

Fig. 4.9 Former warehouse partly repaired and reoccupied as a result of action following the 'Buildings at Risk' pilot study in the Kirklees Borough Council area.

4. Three adjoining terraced houses, devoid of roof structures and in split ownership, presented a considerable challenge. As the £10,000 funding was far from sufficient to deal with the problems the Kirklees Historic BPT was approached with a view to acquiring the properties if the council decided to instigate a Repair Notice. But progress on this option was delayed due to uncertainty over the possibility of development on an adjoining site.
5. A redundant former weaving factory was given planning permission and LBC for conversion to two dwellings. After initial enthusiasm by the owner, delay in instituting the scheme led to the serving of an Urgent Repairs Notice to enforce essential repairs to the roof and rainwater goods which were grant aided. Further delay led to the threat of a Repairs Notice which forced the owner to sell the property. The new owner successfully converted the building into two, subsequently occupied, residential units.

From this evidence, the first action under English Heritage's Buildings at Risk programme was a partial success. It identified a number of approaches to solving the problem including: the targeting of grant aid where existing provisions were inadequate; publicising 'worthy cases' for rehabilitation; locating and encouraging owners to undertake repair action; persuading owners to sell their properties if they were unable or unwilling to take action; assisting in the process of finding a new owner and appropriate new use; involving a BPT; and taking or threatening repair action under statutory powers as a last resort. This necessarily required a commitment from the council in terms of providing additional finance and staffing for the project to find appropriate solutions, including a willingness to use repair powers which many LPAs have been reluctant to do.

The action from the pilot programme was further developed following the national survey of listed buildings. For example, Newcastle City Council, which conducted a survey of its listed buildings in 1990, devised an action programme, which commenced in 1991, to deal with buildings at risk as the first priority within a much wider conservation strategy (Figs 4.10 and 4.11).

A number of buildings found to be at risk were within the council's ownership and thus presented an opportunity to provide exemplar cases if action could be taken. For example, the former Ouseburn School, an unusual grade II listed building in 'pagoda' style, was converted into a major business development centre (Fig. 4.12). The council had recognised the prestige potential of the building being located in an area designated for regeneration and carried out essential repairs at a cost of £250,000. A further £1.3 million funding was secured from the Department of the Environment and the European Community to convert the premises to provide an attractive environment for new business opportunities. It now forms a striking landmark, contributing to the improvement of the surrounding area. By the spring of 1994 the centre was 90 per cent let with 159 new companies being

Fig. 4.10 2–8 Grey Street (and 21 Mosely Street to the rear), Newcastle upon Tyne. 'At risk' due to vacancy for over 20 years. Lack of maintenance has caused the interior to be ruined by water penetration and a dry rot outbreak and the external ashlar façade is also in a poor condition.

able to take advantage of a phased rental scheme and central business support facilities.

Apart from this action the council higlighted the worst 16 buildings (in risk categories 1 and 2) as priority cases for direct action through negotiations with owners to attempt to stabilise the properties. However, with approximately 200 buildings at risk within the city boundary, further indirect action was taken to encourage a climate of care in the city. This included setting up and publishing a register of vacant buildings to encourage new ownership; promotion of the register and the findings of the risk survey;

Fig. 4.11 Nineteenth-century brick warehouses 'at risk' in Broad Chare, Newcastle upon Tyne.

Fig. 4.12 The former Ouseburn School, Newcastle upon Tyne. Identified as being 'at risk' in 1990 but converted into a business development centre in 1993.

advertising the council's willingness to operate policies flexibly to encourage appropriate re-use; publicising the council's willingness to use statutory repair powers; re-targeting available grant moneys towards rescue and preventative work; and lobbying for more grant funding.

After three years 75 buildings had been saved from destruction largely due to the strategy that had been adopted which had encouraged a climate of partnership between many private owners, developers and conservation agencies (City of Newcastle upon Tyne, 1994b). The flexible approach to re-use had allowed some centrally located buildings to be converted to residential use, which had been prevented in the past due to strict car parking standards being applied. Moreover, a number of large buildings were converted to student accommodation, taking advantage of an increase in students in higher education in the city and the guaranteed income from this source. Furthermore, lobbying for further funding led to English Heritage providing substantial funding for particular schemes including a grant of £131,655 to assist the North Housing Association refurbish one particular building at risk.

This latter action is indicative of the recognition by English Heritage of the need to target financial assistance following the findings of the national sample survey of buildings at risk. Moreover, the Newcastle survey had found that over 14 per cent of its listed buildings were at risk, which was twice the national average (Brand, 1992a). The total annual budget of grant aid for 1600 listed buildings and conservation areas within Newcastle upon Tyne was less than £200,000, apart from the limited national pool of grant aid for buildings at risk. This represented a considerable shortfall in terms of the number of buildings in need of assistance. Furthermore, with over 20 per cent of listed buildings in a 'vulnerable' state, few resources remained to support the necessary action to prevent them becoming 'at risk'.

Despite the relative initial success of the strategy employed for risk properties, it was recognised that the underlying problem of economic decline in the historic core of the city required much wider consideration. Of significance, the late-Georgian commercial centre, known as the 'Grainger Town', has a higher than average percentage of 'outstanding' buildings (see Chapter 1), with 20.1 per cent listed grade I and 12.3 per cent listed grade II*, many of which have been under threat due to low rental and capital values resulting in a growing lack of economic confidence. This was caused principally by dramatic changes in shopping patterns, environmental erosion and detrimental traffic management measures over the last 20 years, compounded by landlord absenteeism. However, as indicated in English Heritage's sample survey report, historic building grants are unable to counterbalance a problem of such magnitude.

The Newcastle example confirms the need for conservation policies to be more effectively integrated with planning and regeneration policies. Due to the inadequacy of the existing grant system to deal with building neglect caused by economic decline, consideration of this viewpoint by English

Heritage subsequently led to the development of a new initiative (Page, 1992). In April 1994, 14 pilot 'Conservation Area Partnerships' (CAPs) were introduced to direct limited resources on local authority areas considered most in need or where it was considered most could be achieved. Areas were chosen which combined townscape quality with financial, material and social need, the idea being to form a partnership with the local government at district or county level to tackle the problem of listed building neglect (Johnson, 1993b).

Following the development of an 'action plan' by Newcastle City Council for the Grainger Town, the area was chosen as one of the first partnership schemes and awarded an initial sum of £325,000 for the pilot year. With additional city funding commitment of £250,000, it was decided to raise the maximum grant award for repair work from 40 per cent to between 60 and 80 per cent. The aim has been to continue the pilot projects beyond their pilots year where necessary and to extend the CAP initiative to other areas (see Chapter 7). In the case of Newcastle the length of the programme of support beyond the pilot year is five years, with the option to extend for another three years (City of Newcastle upon Tyne, 1994a).

This partnership arrangement has been principally designed to support repair work with the aim of attracting funding from other sources such as the Millennium Fund and English Partnerships to support development works. Specific targets for the six-year programme include: the *rescue* of 120 buildings in a critical condition through offering 'securing' grants, enforcing repair, encouraging sale, or compulsory purchase; the *repair/re-use* in relation to the vacancy in upper floors of 45 buildings; *raising the quality* of the street environment by encouraging improvement to shop fronts, street spaces and other aspects of the historic environment; and *promotion and education* to inform and assist in the regeneration of the Grainger Town (City of Newcastle upon Tyne 1994b).

Further partnership projects were set up in 1995 with the possibility that they may replace 'Town Scheme' arrangements in the future. Thus English Heritage has marked the way forward for England's neglected listed buildings. The future success of this initiative will nevertheless depend upon political support.

The role of building preservation trusts

Background

Since the Civic Trust first advocated the need for BPTs and the government endorsed the idea by the funding arrangement created under the Architectural Heritage Fund (AHF), the number of BPTs registered with the AHF has grown to over 100 in England with more trusts being added to the list each year (Weir, 1993a). As attention has become more focused on neglected historic buildings on a national scale, the potential of BPTs to deal with part of this problem has been given greater consideration. The expertise of BPTs in rescuing often hopeless cases has been recognised by local authorities, with some county councils actually establishing BPTs to work across their administrative areas, and English Heritage, as evidenced by a number of the buildings at risk surveys actually carried out by BPTs and subsequent action programmes by them on targeted buildings. The work of the Derbyshire Historic BPT's Buildings at Risk Unit based in Wirksworth and Kirklees Historic BPT in the pilot programme in the Huddersfield area are just two examples of the important role of BPTs.

In 1989 the Association of Preservation Trusts (APT) was formed to provide a central support system to existing trusts, many of which had worked in isolation on a small scale, often having to resolve similar problems. The APT is composed of a number of area groups (six in England) and a UK committee made up of area representatives and three representatives of the AHF. With this facility problems may be discussed on a broader scale, allowing experiences and expertise to spread throughout the BPT movement. Moreover, practical advice may be given to anyone interested in setting up a trust.

Further support has been given to BPTs from the government. Paragraph 7.13 of PPG 15 encourages LPAs to find new owners for listed buildings in need of repair, giving particular endorsement to their ownership by BPTs which have access to funds to carry out the necessary repairs and can usually sell a property quickly once rehabilitated. In many cases where BPTs step in and undertake sensitive conversion work it is questionable whether any action would otherwise take place. In this context it is now recognised that BPTs may have an important role in saving buildings following compulsory acquisition where insufficient action has been taken on a Repairs Notice (Weir, 1992a). Indeed, in the case of *Rolf v. North Shropshire D.C.* [1988] JPL 103, the Court of Appeal found that a local authority was entitled to make a Compulsory Purchase Order (CPO) on the basis of a proposed immediate sale to another body such as a BPT to secure its preservation.

Undoubtedly with CAP arrangements benefiting from the targeting of financial assistance in the future, BPTs may have an even greater role to play in relation to listed buildings not contained in conservation areas or otherwise if they are not classed as outstanding (grade I and II*) and priority funding cases. At the same time, the importance of BPTs has been recognised by English Heritage in all contexts in undertaking building rescues which commercial developers would not be prepared to take on due to the greater risk as compared to normal ventures or because the overall deficit faced by a scheme is more than a developer is prepared to accept. Conservation grant awards are often central to the viability of a BPT scheme and in this context advice has been provided by English Heritage on how to undertake feasibility studies, an essential prerequisite to funding (Johnson, 1993c).

In line with this, the AHF has provided grant aid for feasibility studies since 1990. Furthermore, to give greater encouragement to the initiation of BPT projects the AHF raised the amount of grant aid towards feasibility studies in 1993 from 50 to 75 per cent with the maximum grant per project remaining at £5000, but a higher grant of £7500 being made available for exceptionally complex or large-scale studies (Weir, 1994). Increased financial support has also been provided by the DoE (now DNH) via capital grants to the AHF. From the standstill of 1992/93 when no capital grant was provided (the first time since 1986/87), £300,000 was allocated in 1993/94 and a commitment made to increase funding by £100,000 and £200,000 in the next two years respectively, subject to the government's Public Expenditure Scrutiny review (Weir, 1993b). The additional money has allowed the AHF to fund an administrator for the APT, leaving more time for the AHF's Development Officer to concentrate on activities to increase the number of BPTs.

In the period 1992/93 a record number of loans to BPTs were repaid and with the additional government funding the accumulated fund stood at £7.1 million at the commencement of the operating year 1993/94. The capital fund is made available at all times for low interest loans to BPTs.

Methods of operation

Detailed information on the process of setting up and operating a BPT was provided by the AHF in 1989 in a publication entitled *How to Rescue a Ruin by Setting up a Local Buildings Preservation Trust*, which was followed by the APT's 1992 publication entitled *Guidance Notes for Building Preservation Trusts.* In broad terms the methods of operation and various approaches used by BPTs to 'save' buildings are explained as follows:

Although BPTs are voluntary organisations, in order to protect trust members from being personally responsible if a BPT falls into financial difficulties, they are normally incorporated with limited liability provisions as a company 'limited by guarantee' under the Companies Act 1985. This requires the establishment of a management committee (the equivalent to a Board of Directors in company law) which is responsible for the general control, administration and management of the BPT as a charity. The members of the management committee effectively act as charity trustees and are required to appoint a company secretary who is responsible for the BPT's accounts which must be filed annually with the Registrar of Companies. An unincorporated BPT is less likely to deal with large or complex projects due to the greater financial risks involved.

BPTs must register with the Charity Commission under the Charities Act 1980 in order to qualify for charitable status. The objectives of a trust must be charitable, i.e. non-profit making, the purpose of a BPT being to rescue historic buildings for no monetary gain. Charitable status allows certain reliefs from direct taxation (such as income and corporation tax) but not

VAT unless the works carried out qualify as 'approved alterations' to 'protected' buildings.

The Charity Commission regulates the operation of a BPT in certain ways in that its consent is required prior to mortgaging or selling a property and it must be satisfied that the price realised on sale of a rescued building is the best price that can be reasonably obtained and that steps have been taken to ensure the long-term preservation of a building. This latter requirement may be satisfied by a BPT imposing restrictive covenants on the first purchaser or by requiring each successive purchaser to enter into a Deed of Covenant which may incorporate both positive and restrictive covenants. Alternatively, a BPT may retain the freehold in a building and sell the property by way of a long lease, normally for at least 99 years, which is the most effective way of ensuring that both positive and restrictive covenants can be enforced against a future owner.

Charitable status also requires that the management committee work for a trust on a voluntary basis. Members may, nevertheless, be reimbursed for reasonable expenses incurred and for reasonable fees and professional services if specified in a trust's Memorandum of Association. Other activities to support a BPT that are not in themselves charitable, such as fund-raising, must be undertaken by setting up a separate trading company that will covenant its income to the trust.

There are principally two types of BPT. First, *single project trusts* are formed to tackle a particular building or group of properties. They are often set up by members of a local community who are expressly concerned about the condition and future of a particular building. The majority of these trusts, as well as carrying out rehabilitation work, usually maintain and manage the building when the project is complete.

Two examples of single project trusts are the Chatham Historic Dockyard Trust, which was set up following the closure of the historic dockyards in 1984 to repair and rehabilitate buidings located there, and the Crosby Hall Education Trust. This latter trust was established in 1987 to provide a 'safe haven' for children by restoring and converting Crosby Hall and the associated seventeenth-century farm buildings as a residential education centre. Both of these trusts have obtained loans from the AHF (Weir, 1993a).

However, the majority of BPTs work on a *revolving fund* basis. These trusts are usually associated with a particular area (town, district or county) and operate by acquiring, restoring/rehabilitating and disposing of buildings using the surplus derived from sale proceeds, less any loan repayment, as working capital towards the next project. Loans from the AHF in this context are usually short term, depending on the time taken to complete a project and dispose of the building. In the recessionary years of the early 1990s some BPTs had difficulty in realising sales of completed preservation projects. This has led some BPTs to retain ownership, allowing the choice to let their assets wait until the capital value increases.

Funding of BPT projects may be derived from a variety of sources. English Heritage and local authorities are the principal sources of grant aid and other grant-aiding bodies such has the Rural Development Commission have provided a limited amount of financial assistance. Because of their charitable status, some trusts have attracted funding from other charitable sources such as the Pilgrim and Landmark Trusts and even private benefactors. However, the principal source of funding for BPTs is via the low-interest loan facility offered by the AHF.

The AHF will lend up to 75 per cent of the gross cost of each project up to a maximum of £250,000. In return, all borrowers must offer adequate security, usually in the form of the first charge on the building once a project is completed, or on a repayment guarantee from a bank or local authority. The loan period is normally for two years, but in exceptional cases the AHF may agree to allow another year. The interest rate offered has been maintained at a low level, currently 5 per cent, which, in the year ending March 1991, had benefited BPTs by an estimated £1.5 million from the difference between interest on AHF loans and market rates. Even in times of low inflation when market interest rates are generally low the AHF rate still represents a very competitive rate. However, if a loan is not repaid within the agreed period the rate rises to 3 per cent above bank base rate.

Due to the property recession, shortage of finance has led some BPTs to develop ingenious ways of funding works particularly where the building is large and has a high potential end value and where the works may be phased. An example to illustrate this point may be given in relation to the repair and rehabilitation of the grade I listed Alderman Fenwicks House (see Fig. 6.1), a rare surviving late-seventeenth-century merchant's house grade I listed building with later additions, situated in Newcastle upon Tyne, which has been undertaken by Tyne and Wear Building Preservation Trust Ltd. The city council was forced to buy the freehold interest in the building via a Purchase Notice at a cost of £88,000, half of which was met by the DoE, after other attempts to save the building since 1966 had failed. The building was subsequently acquired by the trust in 1981, on a 125-year lease with the first 25 years rent free followed by a rent of £1000 p.a. for the remaining 100 years.

The first three phases of works were completed by 1992 at a cost of approximately £560,000, two-thirds of which were met by English Heritage and the city council; the remainder was derived from private charitable trusts and other fund-raising activities. The final stages of the work has been estimated to cost up to £1.8 million, of which half is to be sought from grant aid via English Heritage and the European Regional Development Fund, and the remainder from loans via the AHF and a commercial lending institution or the Heritage Lottery Fund derived from the National Lottery receipts. With the current value of the building estimated at 1994 prices to be £900,000, two-thirds of this may be raised via a mortgage from the latter source. Once the project has been completed the trust's preferred option is to retain the freehold interest in order to secure an income from leasing the property as offices which, in turn, may enable the building to be remortgaged as its capital value rises and allow funds to be raised for further projects. Thus by this approach the BPT will be able to move away from the revolving fund basis of financing projects (Jobling, 1994).

As the AHF and APT encourage the formation of more BPTs the need for imaginative solutions to the problem of funding may result in other BPTs developing similar long-term solutions. Indeed, with the support of the AHF and APT, seminars and conferences have been frequently arranged for members of the APT to exchange experiences and develop ideas on a wide range of matters including funding, conservative repair techniques and other matters in association with other conservation bodies.

The achievement of building preservation trusts

The BPT movement has undoubtedly been successful in rescuing, rehabilitating and preserving historic buildings (Figs 4.13, 4.14 and 4.15). Largely through the support of the AHF, which in the period 26 May 1976 to 5 April 1993 granted a total of 257 loans with a value of £14,890,200 (Weir, 1993a), many otherwise condemned and unwanted buildings have been saved. The importance of the work of the AHF and the BPTs was recognised by the government's commitment to increase capital funding to the AHF in 1992 in what the Secetary of State for National Heritage described as 'one of the heritage world's good news stories' (Weir, 1993b).

The majority of preservation projects undertaken by BPTs are small scale and are often associated with grade II buildings situated in small towns or rural locations which may be neither priority cases for grant aid nor attractive commercial propositions for rehabilitation. The rescue of hundreds of such buildings is indicative of the important role of BPTs in resolving part of the national task of dealing with historic building neglect. Yet BPTs have also been successful in dealing with quite large projects. For example, the Derbyshire Historic BPT rehabilitated 57 derelict cottages in Derby at an overall cost in 1980 of £1.25 million (Weir, 1989).

More significantly, the fact that the majority of BPT projects are financially successful provides an exemplar to others as to the approach which perhaps ought to be taken with all historic buildings. For instance, the case of Alderman Fenwick's House reveals the potential to raise necessary funding for repair and restoration works to historic buildings situated in commercial centres which may otherwise be condemned to redevelopment behind a retained façade by private developers, as has been the case of many other listed buildings situated in the central business area of Newcastle upon Tyne and elsewhere.

While a BPT project may be the last resort option for a neglected building, this does not mean to say that conservation principles of repair and rehabilitation are rejected on the grounds of cost. Moreover, the 'preservation' work

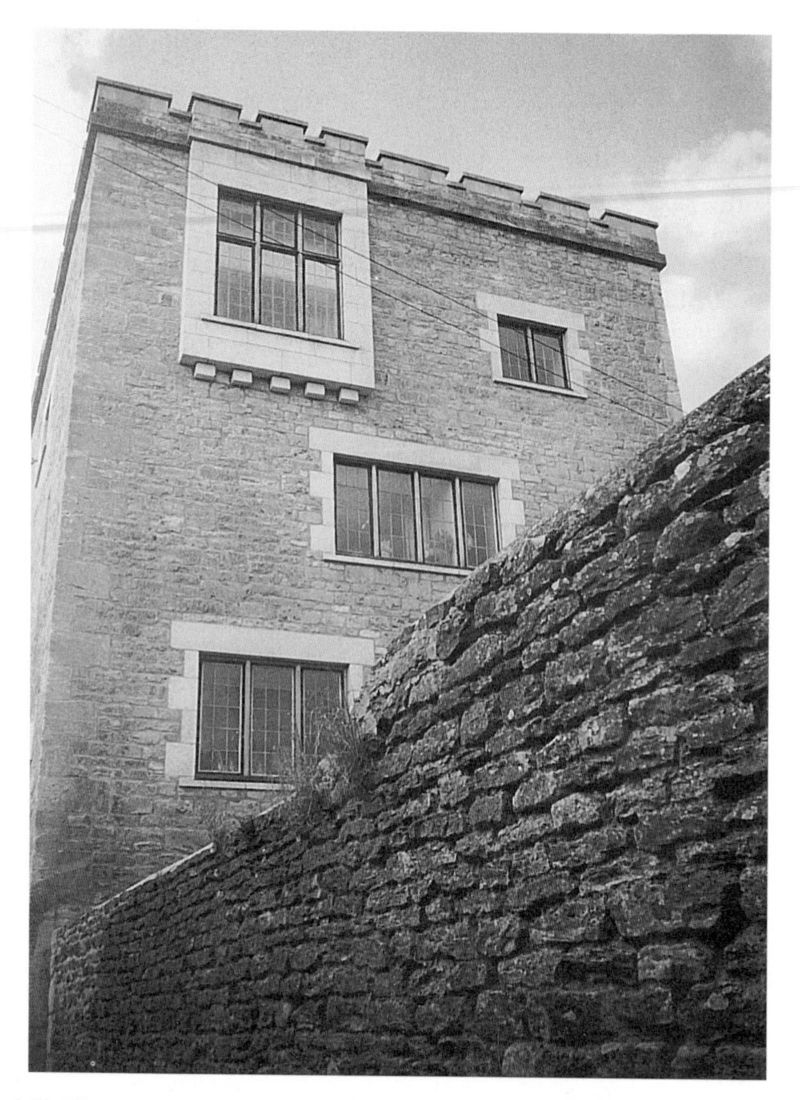

Fig. 4.13 The seventeenth-century Abbey Brewery, Malmesbury, Wiltshire. Formerly 'at risk', but converted to a showroom with gallery space above in 1988–90. This was the first project of Malmesbury Building Preservation Trust. The project benefitted from a grant of £53,000 from English Heritage and a loan of £80,000 from the Architectural Heritage Fund. Other funding was provided by a bank loan. The project provided a cash surplus of £40,000 following the sale of the former Abbey Brewery in 1991 enabling the formulation of a revolving fund for further projects. Malmesbury benefited from the preservation of a unique building and the creation of a trust as an effective alternative to the unsympathetic redevelopments of historic buildings which had previously been carried out in the area.

of BPTs often distinguishes the work of profit-oriented commercial developers. This is frequently because a greater amount of time is required to complete a project than would be acceptable in the commercial world. This,

Fig. 4.14 17 Long Row, Belper, Derbyshire. An exemplary project by the Derbyshire Historic Buildings Trust. The building is one of a block of houses constructed by Jedediah Strutt for his mill workers in the period 1794–7. When the trust purchased the property in 1983 the front wall was bulging and there was a Victorian extension to the rear in need of structural repair. The scheme involved carrying out essential repairs including structural repairs to walls, the roof and existing sash windows in a conservative manner, altering the interior while retaining its character to provide 'essential services' in the form of a bathroom, and reinstating traditional iron railings to create unity with neighbouring properties. A total of £6895 was provided in grant aid from Amber District Council, Derbyshire County Council and English Heritage enabling the property to be sold for a profit of £1055. (English Heritage do not normally reclaim grant aid from Building Preservation Trusts.) The project was used by the trust to illustrate to the local community how an historic building can be repaired and modernised without the loss of original character. Derbyshire Historic Buildings Trust has also published *Fact Sheets for a Restoration Project* in 1992 to illustrate the correct approach in conservation terms to carrying out repairs, alterations and improvements to historic buildings within Derbyshire.

Fig. 4.15 3 Quay Walls, Berwick upon Tweed; built in the late eighteenth century as a granary and warehouse. From about 1960 it was used for storage purposes but fell into disrepair. By the early 1980s the pantile roof had partially collapsed. Berwick upon Tweed Building Preservation Trust implemented a scheme to repair and rehabilate the building involving conversion to nine dwellings, office accommodation and craft workshops with the assistance of a loan of £150,000 from the Architectural Heritage Fund. The conversion was designed to maintain as much as possible of the original fabric and character but also to provide modern accommodation. Externally two lucam windows and a staircase were added to the main elevation. The property was subseqently sold allowing the trust to purchase two further properties.

in turn, allows more time to consider sensitive approaches to building rehabilitation and different forms of fund raising. The fact that a variety of sources of financial assistance are frequently used is not just a measure of need but it is also an indication of the resourcefulness, imagination and patience of BPTs.

Part of the success may be attributed to negotiating a purchase price for a building which truly relects its condition. The AHF has advised BPTs not to be quick to increase the offer for a building sought out for a rescue project where the owner is unwilling to accept the initial offer which has been based on a feasibility study of costs and end use/value. This approach is backed by government policy in that since 1994 the revised conservation policy advice for listed buildings contained in PPG 15 specifies that applicants for LBC to demolish must provide evidence that the property has been offered for sale at a price which fully considers its relative state of repair. This provides an opportunity for BPTs as the new policy begins to bite. Moreover, evidence that some LPAs have used their statutory powers regarding Repairs Notices when suitable cases have been identified by BPTs is an effective way of encouraging buildings on to the market for restoration. The threat of such action may be sufficient to induce an owner to sell at a reasonable price (Kindred, 1992).

Dealing with the reluctant owner

The targeting of finance and strategies for the re-use and repair of builidings at risk is the positive course of action for such buildings. However, if owners cannot be encouraged to take action on buildings then statutory powers may be used to enforce the preservation of listed buildings. This is given official endorsement in s. 7 of PPG 15, although it is generally considered as the final option. But, increasingly, there has been a greater willingness to take action against owners and occupiers of listed buildings who are not prepared to maintain and repair them.

While there are no specific obligations placed on owners of listed buildings to keep them in a good state of repair, the listed building legislation provides a number of measures to prevent the deterioration of the fabric of such buildings. Apart from the more general common law and other statutory powers as may be found, for example, in the Occupiers' Liability Acts of 1954 and 1984, the Building Act 1984 and the Housing Act 1985, the various options under the P(LBCA) Act 1990 may be reviewed (Pickard, 1993c).

Urgent repairs

Under s. 54 of the P(LBCA) Act 1990 a local authority, or English Heritage within London (or elsewhere if authorised by the Secretary of State for

National Heritage), may execute urgent repairs for the preservation of an unoccupied or partly unoccupied listed building. The works are limited to emergency repairs for the purpose of protecting a building from weather damage and vandals and other matters which are thought to be 'urgently necessary for the proper preservation of a building' (Michell, 1988). These will include works to exclude water, to prevent damage from organisms, to ensure safety and stability and damage from fire, etc.

The procedure is commenced via the issuing of an Urgent Repairs Notice giving a minimum of seven days' warning to the owner of an intention to carry out the works which must be clearly described in the notice. The time period is provided to enable the owner to discuss the proposals and consider the merits of undertaking the works or whether there would be grounds to appeal against the notice in accordance with the rules of natural justice.

Under s. 55 of the P(LBCA) Act 1990 expenses to cover the cost of urgent works can be recovered from the 'owner' as defined under s. 336 of the Town and Country Planning Act 1990. There is a right of appeal against such a demand for payment on the basis that the works are: unnecessary; unreasonable; would cause hardship; or, in the case of temporary support works, have continued for an unreasonable length of time. As long as owners are given a realistic amount of time to complete the necessary works, or local authorities are able to defend their costs through obtaining a number of estimates, appeals are unlikely to be pursued. Moreover, unless hardship can be proved it is unlikely that appeals against costs are going to be successful. This view is evidenced by case decisions.

In a case brought by New Forest District Council to obtain reimbursement of £16,922 for urgent repairs to a grade II listed barn and farmhouse at Totton, Hampshire, representations were made on all four grounds of appeal. With the exception of certain minor works the representations were dismissed. The amount recoverable was only slightly reduced to £14,790 (Moore, 1989).

A more significant case may be indicated in relation to the grade I listed Revesby Abbey, Lincolnshire. Following the failure of the owner to comply with a 's. 54 notice', English Heritage carried out works to stem the spread of dry rot, to weatherproof the building, to strengthen the building structurally and make it secure. Subsequently an appeal against costs amounting to £119,000 was dismissed. The case set a clear precedent on costs and a message to owners of the need to secure listed buildings or face the possibility of receiving a large bill later (Kindred, 1991). One further point of interest from this case was that the 'owner' was in fact a tenant for life of a settled estate and was able to benefit from a rarely used procedure contained in s. 87 of the P(LBCA) Act 1990 whereby the Secretary of State granted a certificate which enabled the repairs to be paid for out of capital moneys of the estate rather than the yearly revenue with which a tenant must usually be satisfied (Bird, 1991). This is a significant procedure which may prove to be of substantial benefit to tenants for life of listed buildings forming part of settled estates.

A third case (E1/D3505/4/4/01 [1992] JPL 1190) reveals the strictness of s. 54 procedure time limits placed upon owners. It involved a redundant United Reform Church listed grade II in Nayland, Essex. On the date specified in the notice that the council would move in to do the works, they had only been partly undertaken by the owner's contractor. Following a further site inspection three days later the council's contractors took over the works. This action was allowed by the Secretary of State and the appeal on the ground that the level of costs were unreasonable was dismissed. In fact, due to the failure of the owner to take swift action the cost of the works had increased from the initial quoted sum of £1505 plus VAT to the final sum of £3297. Furthermore, a claim of hardship was dismissed as the property had subsequently been sold for £65,000. Thus the s. 54 procedure can be very effective in securing emergency repairs to listed buildings.

Full repairs

The Repairs Notice procedure under s. 48 of the P(LBCA) Act 1990 is more significant as it can be used on occupied buildings. Furthermore, the Court of Appeal confirmed in the *Rolf* case that there is no right of appeal against recovery of costs on the grounds of hardship. Of greatest significance, failure to carry out the terms of a Repairs Notice within two months can lead to the local authority making a CPO in order to ensure a building's long-term preservation. Indeed, the notice must explain the consequences of non-compliance.

Unfortunately the full repairs procedure has not been very successful in dealing with neglected listed buildings and this is a problem which the government has now recognised (National Audit Office, 1992). In a study covering the six-year period up to the end of March 1990 only 287 Repairs Notices were threatened by councils and in fact only 162 were served on recalcitrant owners (Kindred, 1992). These figures are very low compared to the 36,700 or 7.3 per cent of all listed buildings which were estimated to be at risk by English Heritage in their sample survey (Brand, 1992a). However, while the reason most frequently given for inactivity was fear of the financial implications of CPOs, in fact in only 5 of the 287 cases of threatened or actual Repair Notices were CPOs confirmed. Moreover, it has been reliably suggested that there is on average a less than 1 in 50 chance of a council ending up with a CPO, and in any event there are a number of sympathetic developers and a growing number of BPTs that are usually willing to take on the small number of buildings which are actually acquired (Weir, 1992b).

If anything there has been a new climate for action as evidenced by the 59 LPA areas which took part in English Heritage's buildings at risk sample survey and by the development of the conservation area partnership initiative. Moreover, all LPAs in England were sent a copy of the Association of Conservation Officers research report on the use of Repairs Notices, which exploded the myth that statutory repair action will inevitably lead to a CPO (Kindred, 1992). The more confident LPAs have been prepared to take repairs action, and where they have it has generally been very successful. Certainly the evidence of successful actions will serve as an example to those authorities that have been worried about the perceived threat of CPO proceedings.

Furthermore, official guidance on the historic environment indicated in para. 7.13 of PPG 15 has given encouragement for the use of *'back to back'* deals to resolve the potential financial implications of repair action on serving authorities. These are explained in detail in DoE Circular 11/90 and are set out in Regulations 15–18 of the Local Authorities (Capital Finance) Regulations 1990 (as amended by Regulation 2(d) of the 1991 Regulations). In brief terms, a *back to back* deal is an arrangement whereby the LPA may identify a suitable private individual or body, such as a BPT, who will be willing to purchase a neglected building subject to a Repairs Notice. The building will be purchased by a LPA via a CPO having first entered into a legal agreement with a prospective subsequent purchaser to ensure that a deal will proceed once a CPO has been confirmed (Sutton, 1994).

Finally, on a procedural note it has been confirmed by the House of Lords in *Robbins v. Secretary of State for the Environment* [1989] EGCS 35 that a Repairs Notice should not specify works over and above those which would be reasonably necessary for a building's 'preservation' as opposed to its 'restoration', i.e. a notice can only require a building to be repaired to its condition at the date of listing and not to its former glory at some time before this date. By mounting a challenge in a Magistrates' Court invalid requirements can be disregarded by an owner, but other relevant requirements will remain valid.

Compulsory purchase

The compulsory purchase of a listed building is generally to be regarded as the last resort. This may be evidenced by a case involving the former St Ann's Hotel, part of the grade I listed buildings known as The Cresent, in Buxton, Derbyshire, and designed by John Carr of York in the late eighteenth century. In 1993 the Secretary of State for National Heritage served a CPO on the owners of the buildings, the first time such action has been taken by a Secretary of State rather than a LPA (Baillieu, 1993).

Despite action being taken via two Urgent Repairs Notices in 1990 and 1991 to provide temporary roof cover, carry out urgent works to external stonework and to stem dry rot, the extent of deterioration led to the serving of a Repairs Notice in 1992 on the new owners of the buildings who had bought the premises knowing the situation earlier in the same year. A grant of £225,000 was offered by English Heritage based on a schedule of repairs estimated to cost £900,000, but the owners failed to take any action after three months, one month longer than the required period before a CPO could legally be served. The Secretary of State's involvement was largely due to the high cost of repairs, the need to collaborate on a planning brief for the buildings, and the importance of the building nationally, bearing in mind the responsibility of the new Ministry in this context (Burton, 1993).

In general terms, a historic building owner can avoid CPO proceedings by taking reasonable steps to preserve a building subject to a Repairs Notice. In practice, this is usually what happens as once the date is set for the CPO Inquiry stage most owners realise the seriousness of the LPA's intentions. An owner should not rely on the fact that financially restricted authorities will not be able to afford to purchase a property, as evidenced from the example of 'The Cresent', apart from the fact that under s. 5B of the Historic Buildings and Ancient Monuments Act 1953 English Heritage may decide to defray the expenses incurred. More significantly, if there has been deliberate neglect a LPA may request an order for minimum compensation to be paid under s. 50 of the P(LBCA) Act 1990.

Of the 17 CPO cases which had referred to minimum compensation up to 1990, a final decision by the Secretary of State was only made in six instances. Yet only three of these orders were actually confirmed (Kindred, 1992). In fact para. 7.15 of PPG 15 advises that an order for minimum compensation should only be sought where there is clear evidence that an owner has deliberately allowed a building to fall into disrepair in order to justify its demolition and secure permission for redevelopment of the site. The final outcome of the *Rolf* litigation (in which an application to halt CPO proceedings was rejected by the Court of Appeal) provides an excellent example of this:

The case involved the grade II* listed Pell Wall Hall, an outstanding house designed by Sir John Soane. The owner first applied for listed building consent to demolish the building in 1978 and sought to achieve this aim through various subsequent court hearings. In 1990, following the confirmation of the CPO with a direction that minimum compensation should be paid, the Lands Tribunal set the price for the house and 4.35 acres of land at £1. The LPA was also awarded costs. English Heritage supported the actions of the LPA in what was regarded as a test case (Brand, 1991). In another reported case (Ref. SE3/5273/362/1 [1991] JPL 605; [1991] JPL 1067) concerning a neglected farmhouse in Kent, the direction for minimum compensation was sufficient to induce the owner of the building, the Mid-Kent Water Company, to fully comply with the terms of the Repairs Notice (see also KK/1/16/0/2/WB [1995] JPL 641).

It is significant to reiterate, however, that a direction for minimum compensation can only be made where there has been 'deliberate neglect'. There is evidence that lawyers have advised against claiming minimum compensation unless there has been a clear and proven case of neglect. In two of the successful cases (including Pell Wall Hall) there had previously been an application to demolish the building and evidence of the owner's intransigence. It may also be possible to determine a low value without a minimum compensation direction. An example of this can be found in the Wigan Metropolitan Borough Council area where a listed building was acquired for the nominal sum of £1 due to the high cost of repairs (Kindred, 1992). This is possible because the basis for assessing compensation is according to the normal rules for compulsory purchase, which assumes there is no development potential excepting minor matters as identified in s. 49 of the P(LBCA) Act 1990.

Dangerous structures

Following the outcome of *R. v. Stroud District Council, ex parte Goodernough* [1982] JPL 246, there is now no automatic right to demolish a listed building which is subject to a dangerous structures order. Under s. 56 of the P(LBCA) Act 1990 priority is given to planning legislation and repair action may be considered more appropriate. Moreover, para B.16 of Annex B to PPG 15 advises local authorities that they should not serve a dangerous structures order for a listed building without first contemplating repairs action through their powers under ss. 47, 48 and 54 of the P(LBCA) Act 1990. However, an owner who demolishes a 'dangerous' listed building on the grounds of health and safety without consent can only be prosecuted if he was actually notified of the need for such consent.

Solutions

The solutions to the problem of historic building neglect are many and varied. Moreover, a combination of solutions may be required to obtain the desired result of historic building preservation:

The evidence of Civic Trust and local authority strategies in conjunction with the initiative of English Heritage 'buildings at risk' campaign and CAP proposals have revealed that the causes of historic building neglect are frequently not an isolated building issue. While the campaigning work of bodies such as SAVE Britain's Heritage may have induced other organisations such as BPTs to tackle particular building problems, the need to take a more holistic approach is evident. The CAP initiative in particular should assist in the process of local authorities developing area-based conservation–regeneration strategies as part of the planning framework for areas of recognised architectural and historic importance. Moreover, with the development plan now setting the framework for development decisions the opportunity to attract developers and investors to historic buildings requires, as PPG 15 advises, both positive and negative policies.

This will need cooperation from all sections of the community. Education and promotion may highlight particular problems and raise the profile of historic building preservation. Exemplar preservation projects by BPTs, local authorities and private individuals may be significant in inducing further repair and rehabilitation work. This may also attract a wider audience to the problem, including the local business community and entrepreneurs. Moreover, the tourism potential of historic buildings is well recognised and may be further utilised to make owners aware of the benefits of preserving their historic 'assets'.

The particular problem of building redundancy and obsolescence requires imaginative solutions in the consideration of new uses which should ensure that the special interest in particular buildings is not irretrievably lost. This requires an element of flexibility on the part of LPAs while owners must be prepared to put forward defensible ideas if acceptable solutions are to be realised. The publication of guidance on repairs and alterations, particularly at a local level, may assist in this process as some buildings are easier to re-use than others (see Chapters 5 and 6).

Marketing buildings either before or after preservation work requires special consideration. Much may be achieved by local authorities in publishing local 'buildings at risk' registers, as some already have, and through departments dealing with economic development matters to identify and direct businesss opportunities to vacant buildings. Where owners are encouraged to work with such initiatives there is a greater chance that the building will be offered for sale at a price which truly reflects the state of repair and potential. Further action may be achieved by property agents positively advertising the potential and merits of listed buildings, particularly where an unusual building may have a certain prestige value. This would assist existing initiatives by SAVE Britain's Heritage, which has regularly produced reports on problem buildings and in 1993 published *Bargain Buildings – Historic buildings in need of a new life*, featuring 148 buildings requiring new owners, and SPAB which produces a similar list of buildings in its quarterly journal to members.

The financing of preservation works is probably the greatest hurdle to resolving building neglect. Grant aid and loan facilities are finite. Yet the targeting of grant aid through 'buildings at risk' grants and CAP arrangements will act as an incentive to local authorities to develop action plans to attract funding.

Moreover, if the CAP programme proves successful it is more than likely that this will eventually replace the 'Town Scheme' funding arrangement. However, there will remain a shortfall for certain types of buildings in certain locations. These may be last resort cases which BPTs are expected to rescue; nevertheless, there is no reason why the perseverance shown by BPTs in attracting funding and in developing ingenious solutions to managing finance should not be adopted by other owners of listed buildings.

Even if funds cannot be attracted in the short term, the solution of 'mothballing' a building may be the desired option. In other words, by making a building 'wind and watertight' and by undertaking simple security measures such as closing window and door openings with temporary coverings, the building may be stabilised until cicumstances permit re-use and/or repair work to be undertaken. Alternatively, a temporary use may be contemplated as a short-term measure.

The last resort solution to historic building disrepair is to use statutory powers. The evidence of research on the use of Repairs Notices indicates that the full repairs procedure makes owners face up to problems and is very unlikely to lead to acquisition by CPO. This should give encouragement to LPAs which have feared the consequences of such action. In other respects, rights of entry powers (see Chapter 2) may assist in proving cases of deliberate neglect with the ultimate sanction of compulsory purchase at minimum compensation. The outcome of the Pell Wall Hall case ought to deter this. However, it has been authoritatively argued that the full repairs procedure would benefit from changes in the law such as to make the cost of repairs carried out by a local authority a legal charge upon the property, and removal of the need to prove 'deliberate' neglect could result in more directions to pay minimum compensation for the few cases which reach the CPO stage (Kindred, 1992).

Some commentators have advocated the use of management agreements for listed buildings, similar to the concept of 'planning obligations' provisions found in s. 106 of the Town and Country Planning Act 1990 (as amended), in order to create a more conducive framework for dealing with listed building neglect (Walker, 1994). For instance, an agreement could be devised which defines the area's special interest in a building, indicates what works would or would not require LBC, and is certified by either English Heritage or a LPA conservation officer for a set time period. Every agreement would necessarily have to be individually devised; however, certain factors could be common to all agreements. These could include statements on: the definition of the curtilage, fixtures and fittings; the period of the agreement; aspects which could not be altered without LBC; a requirement to record permitted works; identification of works which should be reversible; provisions for feedback from the agreement; a policing policy; and the names of the parties to the agreement. The benefits of this approach would include: the provision of a fully documented history and a means to record new works allowing the development of a true record of a building's evolution; the avoidance of the idiosyncrasies of local interpretation of procedures in the context of LBC applications for grade II buildings; greater appreciation of the building by the owner as a direct result of commissioning an agreement; a more interactive approach, particularly in the consideration of an owner's needs; a greater chance that a building will become financially viable; and a more realistic basis for future purchasers to consider the degree of possible alteration. However, this is a radical approach which would require considerable research and the development of appropriate statutory provisions before it could have any chance of being adopted (see Appendix 7).

References

Baillieu, A. (1993) Precedent saves hotel's splendour, *The Independent*, 17 April, p. 7.

Bates, E. (1993) The EC funding and culture. Paper presented at *Conservare '93* – The European Heritage Forum held at Ostend, Belgium, 13–17 October.

Binney, M. *et al.* (1979) *Satanic Mills: Industrial Architecture in the Pennines*, SAVE Britain's Heritage.

Binney, M. and Watson-Smyth, M. (1991) The SAVE Britain's Heritage Action Guide, Collins & Brown.

Bird, C. (1991) Recent legal cases, *Conservation Bulletin*, Issue 15, October, pp. 4–5.

Brand, V. (1991) Compulsory compensation and minimum compensation, *Conservation Bulletin*, Issue 13, February, p. 7.

Brand, V. (1992a) *Buildings at Risk: A Sample Survey*, English Heritage.

Brand, V. (1992b) Buildings at risk: the results analysed, *Conservation Bulletin*, Issue 16, February, p. 1.

Burman, P., Pickard, R.D., and Taylor, S. (ed.) (1995) Papers presented at a consultation entitled *The Economics of Architectural Conservation*, held at the Institute of Advanced Architectural Studies, University of York, in association with the European Union Department DG-10, 12–14 February.

Burton, N. (1993) The Cresent, Buxton, Derbyshire: Repairs Notice served, *Conservation Bulletin*, Issue 19, March, pp. 3–4.

Carvill, A. (ed.) (1984) Prosperity through enhancement, *Heritage Outlook*, Vol. 4, No. 4, July/August, pp. 77 and 92.

CIP (1987) *The Calderdale Recipe For Success*, Calderdale M.B.C./Calderdale Inheritance Project.

City of Newcastle upon Tyne (1994a) *Development Bulletin*, Development Department, City of Newcastle upon Tyne, Issue 1, Spring.

City of Newcastle upon Tyne (1994b) *Grainger Town Project, Newcastle upon Tyne: Conservation Area Partnership – Action Plan & Bid*, City of Newcastle upon Tyne/English Heritage/Department of the Environment.

Davey, K. (1992) *Building Conservation Contracts and Grant Aid: A Practical Guide*, E. & F.N. Spon, pp. 99–114 and 182–202.

Davies, P. (1991) The register of buildings at risk in Greater London, *Conservation Bulletin*, Issue 14, June, pp. 5–7.

de la Hay, C. (1993) Knight in shining armour?, *Perspectives on Architecture*, Issue 2, Vol. 1, May, pp. 22–3.

Ekersley, R. (1991) Buildings at risk: historic buildings technical bulletin, *Construction '82*, September, pp. 12–13.

English Heritage (1990) *Directory of Public Sources of Grants for the Repair and Conversion of Historic Buildings*, English Heritage.

Fidler, J. (1987) Buildings at risk: time for action, *Conservation Bulletin*, Issue 1, February, p. 1.

Hudson, N. (1992) The financial consequences of being listed. Conference paper presented to the Historic Houses Association conference on *Planning Procedures And Listed Building Controls*, 15 April.

Jobling, B. (1994) Unpublished information supplied in relation to Tyne and Wear Building Preservation Trust Ltd.

Johnson, S. (1993a) Repair grants 1992–93, *Conservation Bulletin*, Issue 20, July, pp. 8–9.

Johnson, S. (1993b) Conservation area partnerships, *Conservation Bulletin*, Issue 21, November, p. 17.

Johnson, S. (1993c) English Heritage needs assessment and building preservation trusts, *Preservation in Action*, No. 17, pp. 6–7. Architectural Heritage Fund.

Kindred, R. (1991) Pay up if you will not repair, *Context*, No. 31, September, p. 9.

Kindred, R. (1992) *Listed Building Repairs Notices*, Association of Conservation Officers.

Montague of Beaulieu, Lord (1980) *Britain's Historic Buildings: A Policy for their Future Use* (The Montague Report), British Tourist Authority.

MacDonald, M. (1994) Heritage fund tells how to ask for lottery cash, *The Independent*, 18 November, p. 3.

Michell, E. (1988) *Emergency Repairs For Historic Buildings*, English Heritage, p. 13.

Michell, G. (1989a) *Regeneration – New Forms of Community Partnership: Early Lessons*, Civic Trust Regeneration Campaign, Civic Trust, pp. 12–13.
Michell, G. (1989b) *The Wirksworth Story – New Life for an Old Town*, 2nd. edn, Wirksworth Project/Civic Trust.
Moore, V. (ed.) (1989) *Current Topics* [1989] JPL 645.
National Audit Office (1992) *Protecting and Managing England's Heritage Property*. Report by the Comptroller and Auditor-General, HMSO, p. 15.
Nelson, J. (1991) Value Added Tax: a disincentive to sensitive repairs, *Conservation Bulletin*, Issue 14, June, pp. 9–10.
Page, J. (1991) Government increases English Heritage funding, *Conservation Bulletin*, Issue 13, p. 4.
Page, J. (1992) Asking the difficult questions, *Conservation Bulletin*, Issue 18, October, pp. 3–4.
Pearce, D. (1989) *Conservation Today*, Routledge, pp. 67–71.
Pegg, S. (1994) Information supplied from English Heritage's Buildings at Risk Unit.
Pickard, R.D. (1993c) Listed building's: the strengthening of the powers of protection and prevention, *Property Management*, Vol. 11, No. 2, pp. 114–21.
Pickard, R.D. (1993d) Fire safety and protection in historic buildings in England and Ireland – *Structural Survey*, Vol. 12, No. 2, pp. 27–31 and No. 3, pp. 8–11 (1993/94).
RICS/English Heritage/IPD (1993) *The Investment Performance of Listed Buildings*.
Robshaw, P. (1987) Area conservation strategy, *Conservation Bulletin*, Issue 3, October, pp. 1–2.
Sutton, P. (1994) Listed buildings, repairs notices and 'back to back' deals, *Context*, No. 43, September, pp. 7–8.
Taylor, J. (ed.) (1995) *The Building Conservation Directory*, 3rd edn, Catherdral Communications Ltd.
Wade, S. (ed.) (1990) Old farm buildings in a new countryside: redundancy, conservation and conversion in the 1990s. Presented at the proceedings of a conference organised by the Historic Farm Buildings Group in association with the Rural Development Commission, the Royal Institution of Chartered Surveyors and English Heritage, 29 November.
Walker, A. (1994) Conservation: conflict or consent? Paper presented at a conference entitled *Listed Buildings: Economic And Financial Consequences*, University of Cambridge, 23 May.
Weir, H. (1989) *How to Rescue a Ruin by Setting up a Local Buildings Preservation Trust*, Architectural Heritage Fund, p. 13.
Weir, H. (1992a) Repairs notices: major new study shows how powers are actually used, *Preservation in Action*, No. 15, p. 3, Architectural Heritage Fund.
Weir, H. (1992b) *The Architectural Heritage Fund Annual Report 1991–1992.*
Weir, H. (1993a) *The Architectural Heritage Fund Annual Report 1992–1993.*
Weir, H. (1993b) A good news story, *Preservation in Action*, No. 16, p. 1, Architectural Heritage Fund.
Weir, H. (1994) AHF grants for feasibility studies: a new approach, *Preservation in Action*, No. 18, p. 3, Architectural Heritage Fund.

Chapter 5

THE PHILOSOPHY AND PRINCIPLES OF REPAIR AND CONSERVATION

Introduction

In 1991 the British Standard Institution issued a 'draft' *Guide to the Care of Historic Buildings* which emphasised that a conservation policy must be established for each conservation project whether it involves repair or alteration work. It further stated that certain basic tenets of conservation philosophy must be respected to ensure that 'the significant archaeological, historic, social and architectural values' of a particular building fabric and environs are retained. The draft guide was produced in response to a groundswell of opinion that modern building standards, which have been ruled by regulation requirements developed this century, are not necessarily relevant to the repair and re-use of historic buildings constructed before such requirements were implemented.

The idea that historic buildings require sympathetic repair and restoration based on actual knowledge of the history of a building and the materials and techniques first used is not new. The roots of conservation philosophy in England lie in the *anti-scrape* movement of the late nineteenth century. Since this time international opinion through the *International Council of Monuments and Sites* (ICOMOS) has moved to further develop principles of conservative repair and conservation practice. While bodies such as SPAB have fought the rearguard of such principles for many years, official recognition of the sanctity of historic building fabric in England has been slow to develop despite the fact that buildings of special architectural and historic interest have been afforded statutory protection since 1944.

However, since the formation of English Heritage much work has been undertaken to change preconceived ideas that traditional methods of building maintenance are outdated in contrast to modern techniques, while SPAB runs bi-annual specialist repair courses and COTAC provides an umbrella for a developing network of training organisations in architectural conservation and traditional craftsmanship. In 1991 English Heritage published *The Repair of Historic Buildings* by Christoper Brereton (now deceased) which, in addition to the guidance contained in Annex C to PPG 15 on alterations to listed buildings, has been drawn to the attention of LPAs and owners of listed buildings as the standard to be applied to grant-aided projects (Pearcy, 1992).

The philosophy of conservative repair

Historical background

The development of conservation philosophy in England originates from the restoration of ecclesiastical buildings in the late eighteenth and nineteenth centuries. The architect James Wyatt (1746–1813), dubbed 'the Destroyer', was the first of several generations of restorers to be criticised. His 'improvement' work to many decaying cathedrals such as Lichfield, Hereford, Durham, and Salisbury in the last 12 years of the eighteenth century was attacked for being insensitive and destructive, including the removal of decayed external stonework, monuments and screens and creating untrue reconstructions (Brown, 1905).

In the nineteenth century the 'Ecclesiologists' appropriated the ideas of Augustus Pugin, the Gothic rivalist who had advocated the use of the fourteenth century 'decorated' or 'middle pointed' Gothic style as the true English architectural style (Curl, 1990). They favoured the alteration of ancient churches to conform to a hypothetical, ideal form in which perpendicular additions were rebuilt in a style consistent with older parts of a building. However, the Society of Antiquities objected to architects tidying up medieval buildings in order to conform to contemporary ideals (Spence, 1982). Moreover, in his book, *Seven Lamps of Architecture,* John Ruskin, the socialist writer and artist, condemned the work of the restorers in removing decayed surface detail, arguing that the spirit of the dead workman could never be recalled and that the natural effects of the elements created a mysterious interest in what once had been there (Ruskin, 1849).

Despite such criticism much restoration work was carried out in the third quarter of the nineteenth century. But even the restorers began to criticise each other. This can be seen most plainly in relation to the restoration of St Albans Abbey which began in 1856. Sir George Gilbert (then Mr) Scott was commissioned by Lord Grimthorpe and given a largely free rein to save the building from ruin. The first work carried out between 1856 and 1868 was of an urgent preservation character, but a serious conflict arose in 1871 when the architect began to plaster the great central tower. Further work was criticised by Lord Grimthorpe who felt that it had been carried out in a careless manner following a wilful refusal to examine the building properly. Yet most of Scott's work had been to stabilise the building, whereas when Lord Grimthorpe himself assumed responsibility for the works, following Scott's death, the building was transformed. Three large perpendicular buildings from the fifteenth century were removed, as were abutting monastic buildings and the whole of the exterior was remodelled on the lines of an earlier period to conform to the ideal of the 'Ecclesiologists' (Figs 5.1 and 5.2) (Briggs, 1947).

In fact, Scott, in his *Recollections,* considered himself to be a 'conservative restorer', his work at St Albans being mainly confined to preservation in contrast to the destructive restorations of Lord Grimthorpe. He defended his work by commenting that he had 'to combat at once two enemies from either hand' – those who criticised him for not arresting the ruinous state of the Abbey swiftly enough and those who found fault in any restoration work. He further criticised the extreme views of Ruskin against any form of restoration. Yet while he professed to adhere to the importance of arresting decay and avoiding conjectural reinstatements, his other work was less convincing (Briggs, 1947). Most notably, well-meant surface reconstructions at Chester, Lichfield and Worcester cathedrals were severely criticised (Felton and Harvey, 1950).

To some extent Scott blamed unnecessary alterations on 'over-zealous' clergymen, who desired pretentious window replacements, or 'a barbaric builder' or clerk of works, suggesting that the block contract was one of the worst causes of damage working against tentative gradual repair. In 1864 he acknowledged his guilt in his *On the Conservation of Architectural Monuments and Remains* stating that 'we are all of us offenders in this matter' and called for local vigilance committees to watch against over-restoration (Thompson, 1993). But it was Scott's proposals for Tewkesbury Abbey which led William Morris to write to *The Athenaeum* advocating the need for 'an association to be set on foot to protest against all "restoration" that means more than keeping out wind and weather, and, by all means, literary and other, to awaken a feeling that our ancient buildings are not mere ecclesiastical toys, but sacred monuments of the nation's growth and hope' (Briggs, 1947).

Morris founded and became first secretary of the Society for the Protection of Ancient Buildings, commonly known as 'anti-scrape' by reference to the practice of scraping the face of weathered masonry, removing plaster and furnishings, and recreating medieval architectural styles, in 1877. Morris echoed the sentiments of Ruskin that it was … *'no question of expediency or feeling whether we shall preserve buildings of past times or not. We have no right whatever to touch them. They are not ours. They belong, partly to those who built them, and partly to all the generations of mankind who are to follow us'* … in the SPAB *Manifesto* (see Appendix 1) (Thompson, 1993). The main emphasis of the manifesto was to urge architects to … *'stave off decay by daily care, to prop a perilous wall or mend a leaky roof by such means as are obviously meant for support or covering, and show no pretence of other art, and otherwise to resist all tampering with either the fabric or ornament of the building as it stands'* … The preservation of the 'time-worn surface' of a historic building was considered by Morris to be sacrosanct because it provided a historical record of 'the development of man's ideas', it had sentimental and associational value, and it was in itself 'picturesque and beautiful' (Morris, 1884).

The ideas of SPAB were at first dismissed, but after the Society was able to demonstrate practical repair examples it began to attract interest among the

Fig. 5.1 St Albans Abbey before restoration. *(Copyright Richard Dennis Publications.)*

Fig. 5.2 St Albans Abbey as remodelled by Lord Grimthorpe. (*RCHME Crown Copyright.*)

younger architects of the late nineteenth century. This was largely due to the repair work of Philip Webb, one of only three architects on the original SPAB committee of 56 members, who extended Morris's ideas by stating that it should be possible to extend old domestic buildings as well as repairing them. Thus it would be possible to make them more habitable and thereby increase their life expectancy. However, in the spirit of the manifesto, Webb stipulated that such work should be 'plain and unostentatious' and 'harmonise' rather than 'imitate' the earlier work so that, in time, it would tell its own tale and not hide when it was undertaken (Spence, 1982).

The developing influence of the 'conservative repair' philosophy

The influence of the Society spread under Morris and Webb and later through other architects, most notably William Letherby and later Albert Powys. Letherby became a committee member of SPAB in 1893 and in 1906 was appointed surveyor to the previously 'restored' Westminster Abbey, subsequently receiving press attention for successfully carrying out preservation work to the decaying stone fabric through lime-washing. A scholarship, introduced by the Society in 1930, was commemorated to Letherby after his death in the following year, and has been provided ever since to train scholars in the techniques of conservative repair.

Powys, Secretary to SPAB from 1911 to 1936, was a prolific writer on repair work, particularly in relation to cottages and farm buildings. His 1903 *Notes on Repair of Ancient Buildings* were printed in book form in 1929, the validity of which has been proved by the test of time as SPAB subsequently reprinted the book in 1981 and in 1995, recognising that while methods had evolved there was no better summary of the Society's approach to repair work on historic buildings (Powys, 1995).

Of significance, as the world's first national society concerned with the repair of the built heritage, it is not surprising that SPAB was also influential abroad – particularly in relation to countries forming part of the British Empire such as India, Eygpt and Palestine – and also in Europe. By example, Morris coordinated protests over proposed restoration work to St Mark's Church, Venice, and The Great Mosque in Jerusalem and later, through Morris's successor, Thackeray Turner, a successful protest was made against the rebuilding of Heidelberg Castle, Germany (SPAB, 1982).

The wider influence of the Society certainly assisted the process of changing attitudes away from restoration in favour of conservative repair on an international level. The manifesto was a first attempt to set down basic ideals which guided other bodies in developing the philosophy of conservation. The International Museums Office made the first systematic attempt to develop an international code of practice at the *Athens Conference* in 1931. Furthermore, a congress of architects and other specialists associated with historic buildings met in Paris in 1957, and at their second meeting in

Venice in 1964 approved the International Charter for the Conservation of Monuments and Sites, commonly known as the *Venice Charter*, which is an authoritative document of international significance which largely contains and further develops the underlying principles of the SPAB manifesto. ICOMOS was founded in 1965 following the adoption of the *Venice Charter* (see Appendix 2) and still constitutes the only international, non-governmental organisation set up to promote the theory and practice of conservation relating to historic monuments (including buildings), sites and areas.

In recognition of the fact that cultural values must differ between countries, further documents have been produced to supplement the articles of the *Venice Charter* according to special local needs. The first country to do this was Australia, through the Australian ICOMOS branch, which adopted the *Burra Charter* in 1981, although other charters have been adopted on more wide-ranging conservation matters (see Appendix 3). Such documents are important in influencing further development of conservation principles as may be determined to be significant in individual countries. This may be evidenced by the *Charter for the Conservation of Places of Cultural Heritage Value* adopted by New Zealand ICOMOS in 1993, whereas in England the philosophy of conservative repair is largely dealt with through a number of publications from interested bodies.

The continuing debate – restoration or repair?

The disastrous fire at Windsor Castle in 1992 brought the debate on whether it is better to restore, or repair and provide new sympathetic replacement designs to replace lost features into the current arena. The widespread destruction of St George's Hall, particularly the roof and ornate plaster ceiling comprising 624 shields of arms of the 'Order of the Garter', provided an opportunity to make a 'new' building within.

SPAB argued that because Sir Jeffry Wyatville had remodelled the hall in the early nineteenth century in a theatrical medieval style, disguising his predecessors' work in a way that did not deceive anyone, it would be incorrect to go beyond this work as little evidence remained of what he obliterated. Certainly English Heritage did not favour a conjectural restoration. Thus the opportunity was presented to maintain the evocative castellated outline of the castle, yet replace what had been destroyed in a way in which new work would be clearly identifiable (Darley, 1992).

The need for a replacement roof to prevent further damage to the interior was an urgent requirement. In 1993 the external elements of the roof of St George's Hall were restored to the same profile as existed before the fire. The decision was made to allow time for discussion on the approach to be taken on the interior with the possibility of creating new designs for both the hall and adjoining private chapel. Furthermore, a meticulous archaeological investigation commenced immediately after the fire (Thorneycroft,

1993) so that significant parts of the interior could be, in English Heritage's view, 'accurately repaired' (English Heritage/RICS, 1994). Thus a compromise solution was devised by allowing the outcome of the archaeological investigation of the fire debris and existing documentary evidence to determine the approach to be followed. As such, the ceiling could be restored in favour of a new design.

Windsor Castle is not the only recent case of a major historic building in England being restored following fire damage. Despite SPAB's standpoint in the case of grade I listed Uppark House (*c.* 1690), the National Trust embarked on a meticulous restoration of a 'like-for-like' philosophy. A massive salvage operation commenced after the fire, enabled by the fact that a management plan had been devised for this purpose. While the roof had completely collapsed and much of the rest of the building was destroyed, it was decided to 'reconstruct' the building. Wherever possible original material was reinstated following the analysis of debris collected and deposited in dustbins, each representing a grid square of the house, and otherwise replacement work was carried out.

This principle, applied througout the works, was regarded by the restorers as an undoubted success, though achieving accuracy in replicating both external and internal features was admitted to be a 'considerable challenge' (Thorneycroft and Lewis, 1992). While an attempt was made to satisfy the qualms of those that favoured the 'repair' rather than the 'restore' approach, by sending some of the site managers on the SPAB conservative repair course, the pressure of costs arising from an obligation to seek competitive tenders meant that pure repair work had to be compromised to satisfy the insurance cover (Spring, 1991).

The approach taken on both of these two important buildings may, from the viewpoint of 'purists', be regarded as mere 'forgery', not allowing the change to be 'wrought in the unmistakable fashion of the time', to quote the SPAB manifesto. If indeed the ceiling of St George's Hall is restored, will the intention be to deceive the viewer? Morris's view was that if a restoration is not honest the aim must be to deceive. Yet does it actually deceive and is it wrong to do so?

From a philosophical viewpoint it is possible to consider whether it is wrong to replace the old with new in order to deceive the eye. Thus, in the cases of Uppark House and Windsor Castle, which are or will be open to public view, it may be argued that the intention is to allow the public to see the grandeur of the original historical design. Therefore it may be possible to return to a finish which, to the uninformed eye, appears original. This then begs the question whether something is allowed to be recreated for the tourist who is too unfamiliar to be critical. But should appreciation be just a scholarly activity or for those that are otherwise informed? One answer may be that the uninformed viewer should be educated to be able to perceive what is fake and what is not. Otherwise the truth lies hidden and the viewer is corrupted. Moreover, there is the floodgate argument that if change is not

shown to be change, by alteration in sympathetic but modern desig
est repair, then the ability to distinguish change over time will be a
(Wilsmore, 1994).

The very reason why the 'special character' of listed buildings is n defined (see Chapter 1) complies with this argument. Changes which have been made to a historic building over the course of time may be officially recognised to form part of the special character of the whole building, including what is and what is not known about the building. This, however, may only be determined when works are proposed to a building when the merits of new work have to be considered (or otherwise when research activity is pursued to investigate a building). If the statutory protection system necessitates the consideration of special character when change is proposed it should not be limited to work of a non-restorative nature. At the same time, if the control procedures are to insist that only honest repairs are carried out, over the course of time the process of continually undertaking honest repairs may actually damage the special character of a historic building. If the process of control is discretionary it should allow for consideration of both approaches. In this respect both the Venice and Burra charters indicate an approach which appears to have influenced English Heritage's *Principles of Repair* in a way which may be differentiated from the desired objectives of SPAB, as set out in their document entitled *The Purpose of the SPAB* (see below).

Article 9 of the *Venice Charter* acknowledges the process of restoration so long as it is based on respect for *original materials* and authentic documentary evidence. It further states that any restoration work carried out must not be conjectural. Yet it does not clearly explain what is meant by the word 'conjectural'. An inferred meaning may be 'speculative'. Despite its acceptance of a form of restoration, the charter nevertheless infers that such work should be honest, showing a 'contemporary stamp' and by this approach should harmonise with the whole building. Thus a compromise may be achieved where evidence allows this approach to be utilised, allowing contributions of all periods to be respected. The only problem with this approach is that it cannot be clear what is meant by the term *original materials*. The approach adopted by Morris, as indicated in the SPAB manifesto, was that every change to a building over the centuries 'left history ... and was alive with the spirit of deeds done midst its fashioning'. From this it may be considered that the term *original* should be interpreted as meaning anything comprised in the history of a building at any given point in time. If this is correct then it must be doubted whether the approach of the *Venice Charter* will safeguard the contributions of all periods.

The *Burra Charter*, although only specific to Australia, elaborates upon the limited arguments put in favour of restoration in a way which may be relevant elsewhere. Article 13 of the charter identifies restoration as being appropriate if 'there is sufficient evidence of an earlier state of the *fabric* and only if returning the *fabric* to that state recovers the *cultural significance* of the *place*'. Yet the charter does not define what is meant by the term *cultural significance*.

d in national terms, as culture is most usually recog-
, then this may provide a philosophically acceptable
he thorny problem surrounding the concept of 'rest-
problem becomes more complex if a wide-ranging
nificance has to be determined. This may be the case
r European Union countries since the European
per on the Urban Environment, published in 1990,
eate a 'Community system of recognition of the historic and cultural significance of individual buildings and parts of urban areas'. In 1995 further proposals were devised for establishing a 'Community Action Programme' in the field of cultural heritage (the *Raphael* programme); however, if implemented, this will take a considerable amount of time to accomplish (Commission of the European Communities, 1995).

Leaving aside the issue of closer European integration, the *Burra Charter* development may be considered to have relevance. If buildings such as Windsor Castle and Uppark House are as culturally significant to England as the Acropolis is to Greece, then restoration may be argued to be acceptable in limited circumstances. At the same time it will be extremely difficult to establish ground rules for *cultural significance* which is essentially an intangible matter. On the one hand, permitting this approach in so-called isolated situations may be the 'thin end of the wedge' leading in time to the 'floodgate' argument resulting. On the other, restoration may be acceptable in limited situations such as where a fire has destroyed a building of recognised national importance. Thus a compromise may sensibly be achieved. This seems to be the desire of official statements on conservation philosophy.

Principles of repair and conservation

Guidance and education

In an offical capacity, English Heritage, as the statutory adviser to the government on conservation matters, produced a guidance leaflet entitled *Principles of Repair* in 1989 as a prelude to Brereton's authoritative *The Repair of Historic Buildings* as well as contributing to the advice which now forms part of PPG 15, while the British Standards Institution *Guide to the Care of Historic Buildings* remains in draft form (see Appendix 7). SPAB's guidance leaflet *The Purpose of the SPAB* sets out the principles by which it is guided today (based on the original manifesto) emphasising its continuing role in 'constantly studying, developing and improving ways of putting [its] policy into practice through ... advice, teaching and casework' and thus distinguishing it from other conservation societies.

Moreover, SPAB remains a leading authority on how to repair and maintain historic buildings and runs specialist repair courses for craft workers and building professionals and has produced a number of *Technical Pamphlets* and *Information Sheets* on conservative repair and technical procedures relevant to historic buildings. The importance of the latter is recognised outside England with, for example, the Monumentenwacht organisations in the Netherlands and the Flanders region of Belgium utilising such documents in developing maintenance strategies for buildings within their remit. Furthermore, the BPT movement and SPAB forged closer links in 1993 with proposals to conduct joint seminars on the principles of conservation and repair, and joint action on rescue projects (Venning and Weir, 1993).

Apart from the repair courses run by SPAB and initiatives by the Architectural Association in London and the Institute of Advanced Architectural Studies in York, there has been, until recently, a shortfall in specialist training provisions in the philosophy and practice of conservative repair. Largely through the initiatives of ICOMOS and, in England, COTAC this position is gradually changing. There are now a number of universities and colleges running courses in traditional building skills and repair and conservation work. The Royal Institute of Chartered Surveyors developed a 'Building Conservation Group' in 1987 and, more recently, a 'Conservation Skills Panel'. COTAC has actively encouraged the development of a network of regional training centres in conservation work. One such centre in the south of England is the Joint Centre for Heritage Conservation and Management, which was formed in 1990, incorporating the Department of Conservation Sciences at Bournemouth University, the Weal and Downland Open Air Museum at Singleton, the Lime Centre at Morestead, the Centre for the Conservation of the Built Environment at Bursledon and, since 1993, English Heritage's Building Conservation Training Workshop at Fort Brockhurst.

Much has changed since Ruskin and Morris argued against restoration and purported the view that 'machines demeaned the dignity of labour'. The climate now is encouraging the revival of traditional building skills and the production of traditional building materials. Moreover, the publication of an annual *Building Conservation Directory* since 1993 provides evidence of the growing number of specialist suppliers, consultants and craftsmen now involved in building conservation (Taylor, 1995). Of necessity, the precepts of conservation philosophy have developed since the SPAB manifesto with greater debate over the last century, particularly as the scale of the disrepair and maintenance needs now extend far beyond our ancient churches. More often than not it is a matter of achieving a defensible solution, with buildings requiring alterations to provide new life in order to safeguard their long-term preservation.

The basic tenets

While the *Venice Charter* sets a regulatory tone, emphasising, in a similar way to the SPAB manifesto, the common responsibility to safeguard the built heritage for future generations, it is not a legally binding document. It does not have the official status that is afforded to guidance produced by English Heritage. As such it is debatable whether building practitioners will regularly turn to it for guidance. Nevertheless, it is a document of ethics, and has been regularly mentioned in *Context,* the quarterly journal of the now growing and influential *Association of Conservation Officers* (ACO) (Golding, 1994). As more conservation officers are employed and undertake specialist training in conservative repair techniques, knowledge of and recourse to the charter may become more widespread. Furthermore, the fact that all the guidance notes are written as 'guidance' rather than strict rules is representative of the continuing philosophical debate. For this reason it is easy to understand why the 'draft' British Standard, *Guide to the Care of Historic Buildings,* remains in draft form. Yet by reference to the different guidance documents it is possible to indicate the basic tenets of conservation and repair relevant to historic buildings today:

Repair objectives

No building can withstand the effects of neglect and consequent decay. Thus the primary purpose of repair is to restrain the process of decay caused by the effects of nature in a manner which does not damage the special architectural or historic character of a building. The objective of repair must be to preserve a building. Stabilising the historic fabric of a building may then allow for long-term use either for the purpose for which it was designed or for a new use which may be appropriate to its character.

Avoid damaging the building

Historic buildings should not be used for experimentation nor should any more work be carried out than is absolutely necessary to secure the future of a building. 'Minimum intervention' is a key principle in conservation work. Thus no repair work should encroach on the original fabric in a manner which diminishes the authenticity of a building. Selective repair work should be carried out over time wherever possible through maintenance rather than full-scale treatment at one point in time (Figs 5.3, 5.4 and 5.5).

Research and recording

No repair work should be carried out until a full understanding of a building's history has been achieved (Fig. 5.6). This should include an examination of a building's particular architectural and historic qualities, of any repairs and alterations which have been carried out over time, and of the history of its uses. Such investigation may be relevant in interpreting why decay has set in and how it may best be resolved.

Fig. 5.3 Newcastle Castle with stonework dating from the twelth century onwards. Natural wear of the external surface has required remedial work. While the stone replacement is geologically correct and appropriately set in the original line, the general approach in a historic buildings should be to replace only the essential minimum that is required at any one time. Here the 'honest' replacement will nevertheless weather over time. Attempts to artificially distress new stone should be avoided.

The consideration of survey drawings, showing original or previous work; written descriptions, published or unpublished; photographs and illustrations; and documented information services such as the County Sites and Monuments Record, local government records for planning and building

Fig. 5.4 Berwick Bridge, Berwick upon Tweed; *c.* 1626. Where the approach of 'minimum intervention' has been adopted for the replacement of worn masonry.

Fig. 5.5 Dry rot treatment at the servant's quarters at Belsay Hall, Northumberland, which has led to the removal of all the interior plasterwork and timberwork extending through four floors. It is rare that such drastic treatment is necessary to resolve damage caused by an outbreak of dry rot if the cause of moisture ingress is resolved and a building is allowed to dry out.

regulation control or the National Buildings Record held by the RCHME may be helpful. Preliminary archaeological or architectural investigations should be followed by on-site interpretation of a particular structure.

Fig. 5.6 Derwentcote Steel Furnace, County Durham. Recognised to be the only authentic complete cementation furnace in Britain. Archaeologists had recognised the importance of the site since the 1960s. The ruined site was taken into care by English Heritage in 1985. Conservation works began in 1987 commencing with a programme of detailed research and recording using rectified photography, computer photogrammetry and written description. The interior of the buildings were further excavated to provide technological information regarding the site. The fabric was consolidated and re-roofed. The site was subsequently opened to the public in 1991.

A record of a building should be made prior to the commencement of repair work by means of measured drawings and photographs as a guide to designing appropriate remedial work. Furthermore, all new work should be recorded during the works, which, over time, should allow for a cumulative record to build up to assist in the process of carrying out future maintenance and repair work. Satisfactory arrangements should be made for depositing records either with the owner or with the county or national archive facilities. The latter may be particulary relevant where a change of ownership occurs.

It should also be noted that ICOMOS UK published a comprehensive *Manual on Research and Recording Historic Buildings* in 1990 as a guide to the process of researching and recording.

Analysis of condition

Before the design and specification of repairs can be determined it is essential to fully assess the nature and condition of the structure and building materials and the causes of defects to these factors in a historic building. Existing reports from periodic inspections or other professional surveys may give a general assessment on condition but should not be relied upon for detailed work. Any exploratory investigation of defects should be carried out with the greatest of care in order to prevent further damage (Fig. 5.7).

Approach to repairs

Any repair work should maintain the integrity of a building, respecting its age and character (Fig. 5.8). With appropriate treatment it is often unnecessary to replace large sections of materials which are partially suffering from decay. For example, it may be possible to repair structural timbers or even window frames with new sections and often at a saving of considerable expense as compared to replacement work. Where new work is to be undertaken it should fit with the old (and not the other way round) such as by piecing in timber repairs by scarf joints so that more of the historic fabric remains extant (Fig. 5.9). Carrying out repairs *in-situ* will reduce the need for dismantling and assist in maintaining the authentic qualities of a building (Fig. 5.10).

Repairs should be carried out in a simple and 'honest' manner, matching existing materials and methods of construction or by accepted techniques appropriate to historic buildings. Examples of the latter are the use of 'sacrifical' lime-wash coatings which may be renewed, and 'tile dentistry' in decayed stonework first advocated in the early casework of SPAB. Moreover, past repairs may form part of the intrinsic history and character of a building. Their replacement simply for cosmetic reasons may create a 'forgery', particularly if new work is artificially distressed; and, in

Fig. 5.7 Knowle Hill, Tucknall, Derbyshire. This eighteenth-century former stables and cottage associated with a summer house became derelict after their last occupation ceased in 1958. Since this time the buildings suffered foundation failure, other serious structural defects caused by groundwater penetration and vandal attack. It was purchased by the Landmark Trust in 1989. Works to restore the building commenced with the installation of supporting scaffolding, followed by underpinning and buttressing to prevent further damage to the buildings before other conservation works could be carried out.

Fig. 5.8 Traditional tuck pointing. Thin lines of pure white lime putty set in grooves cut in the mortar of the brick joints, contrasted with poor quality over-pointed cement-rich mortar pointing.

Fig. 5.9 Structural repairs to the dome and cupola at the Cathedral Church of St Philip, Birmingham, using green oak, which has a capacity to bend, and retaining the original oak where possible.

Fig. 5.10 Former chapel of St Michael, Saltisford, Warwickshire; c. fifteenth century. Scheduled monument status was removed to enable the local authority to serve a Repairs Notice under listed building legislation. The original window moulding has been used to cut replacement stone to match while the majority of surrounding masonry has been left in place rather than dismantling the whole wall.

any event, if no damage is caused by such repairs there is no reason to replace them.

Langley Gatehouse, Shropshire is the only surviving part of Langley Hall erected *c.* 1610. Archaeological and archival research indicated that the gatehouse was built in one phase incorporating an earlier building. A decision to take the building into the guardianship of English Heritage was reversed. Following this English Heritage approached the Landmark Trust with an offer of 100 per cent grant aid for repairs if they were to convert the building to holiday accommodation. A meticulous measured survey was carried out and archaeologists from the Shropshire Archaeology Unit were closely involved throughout the project. Work commenced in 1922 under the supervision of a SPAB Scholar architect (Thomas, 1993). (See Figs 5.11–5.15.)

New materials and methods of repair will only merit consideration if they have proved themselves by time and if the benefits of their use outweigh any harm that may be caused to the character of a building (Fig. 5.16). Such work should complement but not attempt to copy the existing fabric, so that in time a distinguishable repair may represent another valuable change in a building's history.

Removal of damaging alterations and new additions

The removal of additions may be acceptable but only where they play no significant role in the cumulative architectural or historic interest in a building. For example, a mid-twentieth-century addition to an early nineteenth-century building erected in a compromising manner before the building was listed may merit removal, whereas additions to a Norman building of Romanesque style made in the thirteenth century in Gothic style will form part of the overall interest in a building.

New additions to listed buildings must be considered in the normal process of control, as discussed in Chapter 2. The essential prerequisite to granting relevant consents is that the special character of a building is not jeopardised. Nevertheless, economic grounds may allow substantial change, even including the extreme case of façadism. Moreover, building and fire safety requirements may necessitate damaging alterations. This should be avoided wherever possible and since the revision of the Building Regulations in 1991 the opportunity has been provided to counter the worst effects of fire safety requirements for new uses in buildings of 'special architectural or historic interest' (see Chapter 6). Even so this will require negotiation with the regulators.

Repair and alteration work should not prevent the future re-evaluation of a building. Thus the concept of 'reversibility' may be an important factor to consider where, for instance, a new use is required to rescue a building at risk. The possibility of reversal should not permit further damage to the original fabric (Fig. 5.17).

Fig. 5.11 The main chimney corbel was rebuilt and badly eroded bricks were replaced within the stack. The bedding mix used was 1 white cement : 4 putty-lime : 12 sea-sand, and the pointing mix was 1 white cement : 6 lime putty : 1 sea-sand : 1 brick dust.

Restoration/reconstruction work

While SPAB continues to argue against the reproduction of 'worn or missing parts' even where there is archaeological evidence for their replacement (except in the case of very small-scale items), official advice accepts the view that particular items which are crucial elements of the original design or of

Fig. 5.12 Timber repairs were carried out using traditional techniques without the use of glue or resins, retaining the original oak members wherever possible. Dry oak was generally used for small repairs and new framing members around window openings while green oak was used for the majority of new frame members. The timber frame panels were infilled to match the original plaster found on a surviving panel. The plaster was applied to a matrix of riven oak mixed in two coats; the first coat being 1 white cement : 6 lime-putty ; 1 sea-sand : hair and the second 1 white cement : 6 lime-putty : 1 sea-sand : 1 brick dust : hair. Brick infilling skimmed with plaster and lime-washed, which had been used to replace decayed original lath and plaster, was not done in repair works due to English Heritage's preference to leave repaired brickwork.

Fig. 5.13 The traditional SPAB 'honest' repair of 'tile dentistry' used to repair external masonry.

Fig. 5.14 External ashlar and brickwork pointed in a mix of 1 white cement : 6 lime putty : 1 sea sand : 1 brick dust.

Fig. 5.15 Re-roofing with Harnage slate, a calcarious sandstone hung on treated battens, and stone replacement to central chimney. A temporary roof cover was used in the course of repair works.

structural significance may be reinstated provided sufficient evidence exists for accurate replacement. Speculative reconstruction is unlikely to be justifiable (Fig. 5.18).

However, where a building has been demolished, either partly or fully, legal powers are available to enforce restoration. This should, nevertheless, only be carried out where appropriate evidence is available. The dismantling and reassembly of buildings may be permissible. For example, a considerable part of the Central Railway Station in Newcastle was dismantled and reconstructed to allow for works associated with an underground passenger transport system. In other respects the legal precedent of the *Leominster* case suggests that where at least 50 per cent of the original materials are extant, a building may be capable of reconstuction. In any case, where restoration or reconstruction is proposed its relative merits should be considered according to accepted conservation philosophy and policy.

Architectural salvage

While the practice of 'quarrying' old buildings to repair other historic buildings is not new, the merits and ethics of using second-hand materials may be

Fig. 5.16 An inappropriate use of cement-rich mortar which has prevented moisture from evaporating through joints causing the stone to decay and leaving the hard mortar standing proud. As a general principle, repointing should copy original mortar. The correct mix can be deduced from anaylsing the original mortar. Where this is not possible a mix should be chosen which will produce a mortar which is weaker than the stone. The same approach should apply in the case of brickwork.

Fig. 5.17 Belford Hall, Northumberland. The repair and conversion scheme adopted here resulted in the horizontal subdivision of James Paine's 1756 palladian house into apartments. The creation of rooms within rooms on the ground floor safeguarded sections of original ornate plasterwork, which had been riddled by dry rot, within a void between ground and first floor. The protection of surviving mouldings in this way allows scope for the possibility of reversing alterations at some future date.

Fig. 5.18 The preservation of Sutton House, the oldest house in east London, was based on an archaeological survey carried out by English Heritage. Wherever possible the building was repaired on the principle of minimum intervention. A few pieces of restoration were carried out for practical reasons, including the re-opening of a doorway. The materials and techniques of the original Tudor builder were followed to ensure an authentic hand-crafted finish.

questioned. There are now over 100 architectural salvage outlets in England comprising two basic forms. First, some dealers specialise in the trade of architectural antiques such as doorcases, chimneypieces and statues. Secondly, demolition salvage merchants collect re-usable materials such as tiles, bricks and skirting boards. Many of these companies are legitimate traders. The market in second-hand materials, which can have a high resale value due to scarcity, sometimes attracts disreputable traders. The acquisition of materials for repair/reinstatement purposes from such sources may be acceptable as long as other buildings have not been 'cannibalised' legally or illegally.

An argument in favour of this source of traditional materials is that it provides an incentive to repair where otherwise it has proved difficult to obtain the required materials. Yet it is not just the question mark over where and how the materials have been obtained that makes this option debatable, but second-hand materials may be, in themselves, defective, or damaged in transport, or cleaned in an inappropriate manner before sale, or re-used in an inappropriate context.

However, there is now an increasing number of firms specialising in the production of traditional materials which will assist future repair work, as may be evidence in *The Building Conservation Directory* mentioned previously. This practice, with the revival of traditional building crafts and skills, will greatly assist the repair of the nation's historic building stock in the future Fig. 5.19).

Future protection

This last factor is significant to the final principle of repair. In order to safeguard the future of our historic buildings, regular maintenance, according to accepted conservation principles, is the key to their practical and economic preservation.

The principles set out here are just basic principles. The English Heritage publication *The Repair of Historic Buildings* sets out the most appropriate forms of repair for different types of materials and structures. In greater detail there are now a considerable number of specialist publications on such individual matters as masonry, leadwork, and thatching which should be considered in depth for relevant repair work.

References

Brereton, C. (1991) *The Repair of Historic Buildings: Advice on Principles and Methods*, English Heritage.

Briggs, M.S. (1947) *Men of Taste: From Pharaoh to Ruskin*, Batsford, pp. 204–24.

Brown, G.B. (1905) *The Care of Ancient Monuments*, Cambridge University Press.

Commission of the European Communities (1995) *European Community Action In Support of Culture*, COM(95) 110 final.

Curl, J.S. (1990) *Victorian Architecture*, David & Charles, pp. 27–60.

Darley, G. (1992) A time when history should not repeat itself, *The Observer*, 29 November, p. 59.

English Heritage/RICS (1994) *Insuring your Historic Buildings: Houses and Commercial Buildings*. A joint statement by English Heritage and the Royal Institute of Chartered Surveyors.

Felton, H. and Harvey, J. (1950) *English Cathedrals*, Batsford, pp. 69–76.

Golding, F. (1994) The Work Of ICOMOS UK, *Context*, No. 41, March, pp. 24–5.

Morris, W. (1884) Paper to SPAB Annual Meeting, July 1884, reprinted in May Morris's *William Morris: Artist, Writer, Socialist* (1936), Vol. 1, p. 124.

Pearcy, O. (1992) Jeremy Christopher Brereton, *Conservation Bulletin*, Issue 17, June, p. 28.

Powys, A.R. (1995) *Repair of Ancient Buildings*, 3rd edn, SPAB.

Ruskin, J. (1849) *Seven Lamps of Architecture*, Chapter 6, paras 18–19.

Fig. 5.19 Rethatching in Bishopstone, Wiltshire; the building was originally thatched but later re-roofed with another material.

Spence, R. (1982) Theory and practice in the early work of the Society for the Protection of Ancient Monuments, in an exhibition journal entitled *A School of Rational Builders*, Society for the Protection of Ancient Buildings, pp. 5–9.

SPAB (1982) Exhibition journal entitled *A School of Rational Builders*, Society for the Protection of Ancient Buildings, pp. 10–32.

Spring, M. (1991) The caretakers, *Building*, 27 September, pp. 42–6.

Taylor, J. (ed.) (1995) *The Building Conservation Directory 1995*, 3rd edn, Catherdral Communications Limited.

Thomas, A. (1993) The Repair that Almost Foundered, *SPAB News*, Vol. 14, No. 4, pp. 10–13.

Thompson, P. (1993) *The Work of William Morris*, 3rd edn, Oxford, pp. 56–74.

Thorneycroft, J. (1993) Windsor Castle fire, *Conservation Bulletin*, Issue 19, March, p. 32.

Thorneycroft, J. and Lewis, P. (1992) Death and tranfiguration, *Conservation Bulletin*, Issue 18, October, pp. 12–15.

Venning, P. and Weir, H. (1993) A helping hand: the Society for the Protection of Ancient Buildings, *Preservation in Action*, No. 18, Winter 1993/94, pp. 6–7.

Wilsmore, S. (1994) The ethics of restoration – Is it wrong to restore buildings? Paper presented under the theme of *Conservation Philosophy and Practice* at the Institute of Advanced Architectural Studies, University of York, 1 February.

Chapter 6

CONSERVATION PROJECTS

Introduction

Whether a building conservation project is purely concerned with repair or restoration work or also involves adaption work for a new use to secure long-term preservation, it should have a sound philosophical basis so that the work proposed is sensitive to the special interest in a historic building. The history and condition of a building requires careful analysis, as does its capability of acceptable beneficial use if it is unoccupied or redundant, bearing in mind planning and building controls and potential user requirements. This will necessarily involve careful analysis of financial considerations including after value as costs associated with repair and adaption of historic buildings are invariably higher than those concerned with other building projects. Moreover, the choice of contractor and most appropriate form of contract for conservation work requires careful consideration due to the sensitive architectural, historic and possible archaeological factors that may need to be considered.

A building conservation project will require a project team of professionals with expertise in the field of building conservation to manage and supervise the work and market the 'end-product' where necessary. Once a project has been completed, its future maintenance and after care should be assessed and planned out.

Initial project assessment

Leaving aside the problems of defining 'cultural significance', as highlighted in Chapter 5, both the Australian *Burra Charter* (Article 25) and the draft British Standard *Guide to the Care of Historic Buildings* (para. 6.1) indicate the need to establish a 'conservation policy' for every conservation project.

An initial assessment must therefore be made to establish the nature of the special interest in the building. Although the statutory list of buildings provides an indication of the relative importance of a building by the grading system, it must be reiterated that the list description, while providing some evidence of the 'special interest', is not conclusive and is not usually detailed. The special interest may be derived from architectural and historic factors as indicated in the 'Criteria for listing' in Chapter 1 which require detailed examination in each case in terms of a building's materials of construction; plan form; design; history of occupancy, use, alterations and repairs; and its location and setting. The initial assessment must also consider its present state of repair and structural condition and its flexibility to allow a new use where this is required to secure long-term preservation.

These are the essential prerequisites which must be ascertained prior to the development of a strategy for conservation and design solutions for sustainable use. Thus, solutions may be considered which justify any proposed interventions into the historic fabric once a detailed understanding of a building has been achieved. As indicated in Chapter 3, under 'Pre-application discussions and advice', the views of the LPA and any other relevant interested bodies should be sought at the earliest possible stage on proposals and ideas and the likelihood of obtaining grant aid. These will also require an assessment of the financial feasibility of proposals.

Historical research

The purpose of researching a building's history is to know its 'story' so that a philosophically acceptable approach may be determined for a project. It may provide knowledge for structural diagnosis of strengths and, more particularly, weaknesses – for instance, where additions made during the course of a building's development have weakened the structural quality of the original building. It may explain the symptoms of disrepair, including that which has arisen from previous repair or other interventions in the fabric. It may improve knowledge of the architectural, historic or archaeological value of a building thereby allowing greater sensitivity in the design of new repair or alteration work. It may also provide evidence for the philosophical acceptance of restoring missing features (Fig. 6.1). A summary of the research findings by the relevant building professional should also be provided for the client who is proposing the works in order to explain the relevant problems and assure the need for what may be proposed.

However, before attempting to survey a building it may be useful to undertake preliminary research from published and unpublished sources. This may involve gathering evidence from a number of published sources and visiting various information depositories. Information derived from such investigation may assist in developing a course of action for scrutinising the building itself.

There are a number of basic published sources which may be referred to. These include: the county-by-county volumes of the *Buildings of England* series, originally written by Sir Nicholaus Pevsner and presently being updated; the volumes of *Survey of London* series commenced by the London

Fig. 6.1 Alderman Fenwick's House, Newcastle upon Tyne; in the process of restoration by Tyne and Wear Building Preservation Trust. The rear elevation shows windows reinstated based on historical evidence. However, while it is known that there was a cupola to the top of the building it is unclear from existing evidence whether this was surrounded by a ballustrade. The last phase of work to this building will see the resolution of this problem. Opinion differs between the scheme architects and English Heritage, however, as to the approach to the restoration of the top section which has been temporarily covered in bituminous felting.

County Council and presently being updated by the RCHME; the *Victoria County Histories*; the statutory list of buildings of special architectural and historic importance held at LPAs and principal reference libraries; and various dictionaries of British architects and books on building materials, building types, styles of architecture and town, county or other topographical studies, the oldest of which are likely to be out of print and may only be obtainable from antiquarian book dealers. More advanced published sources include studies of individual buildings, including works published by the architect of a particular building or biographies of architects or of the patron who commissioned a building and *Country Life* for large historic Country Houses.

Many unpublished sources may also assist in building up the story of a building's development. Public depositories include newspaper and journal archives; public libraries; County Records Offices; the British Architectural Library of the Royal Institute of British Architects (RIBA); the National Buildings Record held by the RCHME; the Victoria and Albert Museum and The British Museum. It may also be possible to gain access to information maintained in private archives.

Relevant unpublished information about a building may include survey reports; original and alteration plans; contract drawings; photographs; paintings or sketches; building accounts, bills, receipts and orders; documentary evidence of gifts, donations and grants; inventories such as may be made on the death of a building owner or in connection with the commencement and completion of a lease term; diaries; legal documents including title deeds, leases, and wills; other official documents including planning and building control files and permissions, rate books and records, drainage consents, censuses and faculties of jurisdiction in relation to the Church of England; and ancient manuscripts which may indicate the establishment of a building.

This initial investigation should be followed by a reconnaissance visit of a building's environs prior to commencing an on-site inspection of the property itself. Reference to existing or old Ordnance survey or other maps, as well as visual cognisance of the development of a surrounding settlement or other change in the locality may also increase knowledge of a particular building and its site.

The last source of information is the building itself. The form of construction in terms of materials, methods and style of composition provide evidence of the relative importance of a building and its features, including the date of the original works, alterations and repair work. Various clues regarding a building's development may be found in marked dates, carpenters' and masons' marks, manufacture marks on materials and the 'casual graffiti' of craftsmen and other indicators (Feilden, 1994). The presentation of survey findings through a survey report, measured drawings and other survey techniques such as photogrammetry will also assist in the process of understanding a building. In this respect the publication entitled *Measurement and Recording of Historic Buildings* provides an invaluable guide to the process of making a historical record of a building (Swallow *et al.*, 1993).

Improvements in technology are also allowing new approaches to survey and recording work by transferring the fabric of a building on to computer disks creating a 'virtual reality'. This pioneering new approach to surveying historic buildings utilises laser technology and conventional recording methods to create a database of information which may be used by conservators, surveyors, architects and contractors (Wainwight, 1994).

Location

The location of a building is significant in a number of ways to the initial assessment of a conservation project. First, the origins of historic buildings are usually to be found in their surrounding locality in the sense that until the advent of improved transport systems through the development of the canals and railways, buildings were generally constructed in locally found materials and through local craftsmanship. Many old buildings have a sense of place about them and this should be respected in the consideration of new work and repairs. Moreover, the availability of materials and appropriate skills for repair and maintenance should be taken into account.

As the legal protection afforded to listed buildings applies to 'the setting' as well as to the building itself (see Chapter 2), the gardens, grounds, relationship to other adjoining or nearby historic buildings and the wider historic area value should all be considered in the assessment. Of significance, 'the setting' of a listed building is similarly undefined in legal terms as a building's 'special interest', and as PPG 15 specifically indicates the effect of works on 'the setting' as being a criterion for granting LBC, this issue requires scrutiny bearing in mind that each case will be considered on its particular merits.

The often quoted successful ingredients of the property market of 'location, location and location' have particular relevance in relation to historic buildings. Geographical location is a significant factor in the financial viability of a conservation project. The evidence of two important research reports regarding listed buildings confirm this. First, the Investment Property Databank's research into the investment performance of listed buildings in office use over the period 1980–92 concluded that older listed properties outperformed other categories of office property in investment terms in central London where the majority of listed and unlisted office properties are located. However, outside London and the south-east listed buildings in office use failed to keep pace with other categories of office buildings (RICS/English Heritage/IPD, 1993). Thus restrictions on the use of listed buildings, whether perceived or actual, may be said to have had a detrimental impact on economic viability in locations where the capital and rental values and the potential of growth in these factors, i.e. outside London, are at their lowest.

Secondly, the English Heritage 'Buildings at Risk' sample survey confirmed that a historic building's chance of survival is likely to relate to its location.

Almost two-thirds of buildings were found to be 'at risk' were not located in conservation areas. There was also found to be a higher degree of risk in rural areas with agricultural property having the second highest number of 'at risk' properties per category. Furthermore, there were found to be a high proportion of 'at risk' buildings in less affluent urban areas. Listed warehouses and industrial mills being of the greatest concern, particularly in industrial revolution towns whose prospects have since declined (Brand, 1992).

Remote rural locations will obviously limit the scope for finding new economic uses and investors are otherwise less likely to fund a conservation project unless there is scope for showing a financial return. Inevitably some building conservation projects will only be considered by BPTs or other charitable concerns such as the Landmark Trust, which has specialised in converting rural properties into holiday accommodation. In other respects, the availability of funding, including grant aid, will be an important factor. Clearly grade II listed properties located in conservation areas are in a better position to benefit in this respect than those of the same grade in other locations, particularly where a building is located within a CAP-designated area. Furthermore, where a conservation strategy has been set up, a positive policy for conservation may enhance the scope for beneficial use of historic buildings instead of redevelopment, as is evidenced by the Calderdale Inheritance Project highlighted in Chapter 4. Other factors, including traffic intensity and transport communication systems, as well as pollution and security, also have a bearing on location assessment.

Condition

In many cases historic buildings have passed the test of time. The very fact that a building remains standing after many centuries is indicative of the durability of its structure. Major decay is often due to long-term water penetration or fire damage, although unoccupied buildings may have suffered from theft and vandalism. Yet it is often the case that emergency repairs to the roof and water disposal system and other safety measures can stem the main cause of deterioration in a structure. Moreover, simple repairs may be all that are necessary if the original structure is of good quality, with sufficient strength, flexibility, and overall stability that has allowed the building to remain standing for a long period of time.

Nevertheless, the condition of a historic buiding is a fundamental consideration in any conservation project, and can only be ascertained by detailed investigation. Existing condition reports may be a useful guide to the building but cannot always be relied upon as being objective. Much depends on the purpose for which the building was assessed. It is therefore necessary to carry out a detailed survey for the initial assessment in order to be able to devise defensible remedial or alteration work. This requires careful examination of roof coverings, chimneys, rainwater disposal goods, walls, windows and doors, internal finishes, floors and staircases, fittings, and services. Detailed advice on inspection matters and the cause and remedy of particular defects may be sought from experts in the field of study and the various organisations involved in conservation work (see Appendix 4), and specialist publications including *Conservation of Historic Buildings* (Feilden, 1994) *Structural renovations to traditional buildings* (CIRIA, 1986), *Conserving Buildings: A Guide to Techniques and Materials* (Weaver, 1993) and the English Heritage Technical Handbooks entitled *Practical Building Conservation* (Ashurst and Ashurst, 1988) and a number of other publications on particular materials.

As many historic buildings may have been built up over time, with many alterations and additions carried out in different materials, it may also be necessary to assess the structural form in some depth, particularly if a building is to be adapted for a new use. The problem here may be best explained by comparing the different forms of structural stresses in building form. In buildings of masonry construction the loading is largely of a compression nature, whereas with a timber-framed building the structural stresses are largely of a tensile nature. Yet where, for example, a timber-framed building has had additions over time such as a new façade or a chimney of brick construction it is more difficult to be certain how the different forms of construction will interact with each other as there is a mixture of compression and tensile support. Thus, in addition to the services of a conservation architect or conservation surveyor in appraising the condition of a building, it may also be necessary to utilise the services of a structural engineer experienced in dealing with historic buildings. Moreover, where a new use is to be considered a structural appraisal of a building may be necessary as part of the initial assessment of the project.

In appraising a building a number of factors require consideration. Of greatest significance, the relevant strength of a building is likely to be indicated by its recent history up to one hundred years and can be assessed by structural calculations. Other factors which should be considered are the flexibility or stiffness to ground movement and the robustness of a building in terms of its ability to accept abuse and not react in a detrimental manner. For example, many buildings with a high tensile capacity in timber flooring may, despite sagging, remain relatively stable. It is important to investigate the factors that provide the overall stability of a building which is often found in a nebulous way from different parts of the structure reacting with each other (Bowles, 1992; CIRIA, 1986).

A fundamental consideration before designing new work or adapting the building to provide a new use is the question of the original structural quality. This may be considered from determining if there were any building standards at the time, whether the building was constructed in a boom period or not, and the relative quality of workmanship employed (Baxter, 1985). This consideration also applies to subsequent alterations in that it is important to determine whether they may have weakened the structural quality of the original building.

Lastly, any decay in the building should be fully investigated, particularly where its source or extent is hidden, as may be the case where an outbreak of dry rot had occurred behind timber pannelling or ornate plasterwork. This may reduce the need for further interventions into the fabric at some later date. Structural defects may also have occurred due to weaknesses inherent in the ground conditions, which may require research into the history of the site and possibly an archaeological investigation.

Flexibility to use

Official policy on the use of historic buildings is indicated in paras 3.8–3.11 of PPG 15. Paragraph 10 states:

The best use will very often be the use for which the building was designed, and the continuation or reinstatement of that use should certainly be the first option when the future of a building is considered. But not all original uses will now be viable or even necessarily appropriate: the nature of uses change over time, so that in some cases the original use may now be less compatible with the building than an alternative ... Policies for development and listed building controls should recognise the need for flexibility where new uses have to be considered to secure a building's survival.

Thus when the original use is totally impracticable it is possible to consider alternative uses for a building.

The realities of the commercial world have often resulted in the economic exploitation of old buildings yet this course of action almost inevitably results in some material change which is damaging to the integrity of the historic fabric. The most acceptable route to change is one which marries change for economic reasons with a philosophically defendable solution. It is therefore essential to have early discussions with the LPA conservation officer or other relevant planning official, as part of the initial project assessment, who will have the burden of determining the balance between these factors at the first instance (see Chapter 3). Negotiations in this context may need to consider the potential for reversibility in the design of alterations if substantial change is proposed so that the possibility of returning a building to its original use may be kept open should the circumstances permit this at some later date.

In general terms some buildings are more malleable to beneficial re-use than others, but all proposals for re-use are likely to incur some philosophical, practical and economic problems which must be appraised in the initial project assessment. For example, the question of loading requirements for new uses is one which requires particular scrutiny, involving the strengthening of floors and structural joints. Yet perceived requirements by investors are not always necessary for a proposed use (Redman, 1994).

Other practical problems may arise such as the need to upgrade services and, consequently, the need to develop strategies to hide new services within the historic fabric (Preston *et al.*, 1993). Electrical installations in particular should not needlessly damage the building and there may be a need for specialist advice regarding where to locate cabling and the design of fittings so as to have minimum visual impact within interiors (Hunt, 1989). The provision of improved heating systems also present a rehabilitation challenge in terms of avoiding condensation problems or decay from vapours and gases, as do the location of heating plant and associated distribution and dispersal systems and for the means of removing the products of combustion (Feilden, 1994). Investment institution requirements for air conditioning in offices are not really appropriate for the majority of old buildings to be adapted for this use, but provision may nevertheless have to be considered, in particular, for museum uses. The rehabilitation of historic buildings may also present acoustical problems and difficulties arising out of the need for improved sound and thermal insulation. However, pressure for unnecessary change should be resisted if a building is capable of attracting an occupier that will ensure the economic preservation of the building.

A change of use or material alterations to adapt a building for a new use will, nevertheless, result in the need to satisfy official regulations. Apart from the need to obtain LBC, approval must be sought in relation to the building regulations. This can pose a real dilemma for re-use as the provisions, particularly those dealing with fire safety, can impose requirements which may be damaging to the character of a historic building. For instance, internal finishes and components such as timber-pannelled walls and doors could have to be compromised to reach a required fire resistance. However, criticism levelled at the regulations, that they have been designed principally for new buildings, have been given official recognition and the Building Regulations 1991 have since provided more flexibility to design fire precaution measures to suit the individual needs of a building.

Paragraph 0.11 of Approved Document B, guidance note on fire safety, which accompanies the 1991 Regulations, indicates that variations in the provisions may be appropriate in the case of buildings of 'special architectural or historic interest' where compliance might prove unduly restrictive. Instead of an emphasis on 'passive' or structural fire safety arrangements, which may intrude upon the character of a historic building, a greater mixture of 'passive' and 'active' measures is now recommended (Pickard, 1993). Moreover, mandatory rules for means of escape, such as the provision of external staircases which may have a detrimental impact upon a historic building, no longer exist. Furthermore, a 'fire safety engineering' approach may be considered so that fire safety aspects can be negotiated according to the needs of the building as well as the potential occupants. This allows consideration of the possible trade-off between structural interventions and 'active' measures for smoke and fire control, detection, and extinguishment systems as well as management policies which may be provided in a fire safety strategy for a particular building (Pickard, 1994).

In other respects change of use may require building regulation approval, particularly where structural change is envisaged. As para. 3.26 of PPG 15 indicates, sufficient flexibility is provided in the Building Regulations 1991 to have regard to matters of historical or architectural value. Nevertheless, in order to ensure that the issues raised by the regulations which may have an unacceptable impact on a historic building are given due consideration, discussions with the relevant local authority controlling officers will be required from the outset and, in particular, before a LBC application has been determined. Apart from structural and safety considerations further discussions may be required to cater for disabled access in a sensitive manner. Environmental health considerations may also be raised such as damp-proofing and the need for insulation. Thus the initial project assessment should anticipate a wide range of matters which may require negotiation before approval for a proposed new use can be given.

Certain types of buildings may more easily lend themselves to new uses than others. A brief consideration of different property types highlights the problems and potential for re-use:

Ecclesiastical Buildings

The potential loss of church and chapel buildings following redundancy is an emotive issue even among members of the secular society. Yet finding an acceptable new use can pose a daunting task.

Square-shaped, flat-ceilinged, simply designed non-conformist buildings are often relatively easily divided into smaller spaces for single unit use and many have been successfully converted into residential accommodation and other uses without destroying their character, although internal fixtures and fittings are often lost in this process (Saunders, 1991). However, the internal arrangement of areas within Catholic and Church of England buildings, particularly those of Gothic design, with high-trussed or vaulted ceilings, traceried and large celestial windows, a tower or spire, pose a difficult problem in terms of finding an acceptable and beneficial use without destroying internal and external aesthetic value.

While many former churches have been converted to office, residential and retail use, in conservation terms other uses which do not significantly break up the internal area or necessitate roof lights to create adequate levels of natural lighting are more acceptable (SAVE, 1987). Yet community use, galleries, warehouses, concert halls, theatres and other large volume users are rare and are less likely to make a project financially viable. Church buildings can also be notoriously difficult to heat and pose problems in providing services, particularly in remote rural locations, and the necessity of disturbing graveyards in this context will require an Exhumation Order to be obtained. In the case of Church of England buildings a new use will require consent through the Church legal system and will always be vetted for suitability thereby normally ruling out uses such as for gambling or the sale of alcohol.

The most obvious choice for the re-use of a historic church is for another religious occupier, such as a Mosque, which may minimise the need for change. Where a church is of significant quality, perhaps with important original fittings, and cannot be found an acceptable new use, it may, as a last resort, be 'protected by other means' by, for instance, the Historic Churches Preservation Trust or, in the case of Church of England buildings, by vesting a church in the Churches Conservation Trust (formerly the Redundant Churches Fund).

Wind and watermills

The importance of windmills and watermills to our national heritage is evidenced by the fact that the SPAB has had a specific section within the organisation devoted to their conservation and protection since 1946. These are essentially 'machine buildings' in which the working parts were designed with the building and together form part of the 'special interest' when listed. Thus, where the machinery is intact these buildings do not readily lend themselves to adaption. Even if the working parts have been removed windmills in particular are difficult to adapt due to their inherent peculiarities of shape as buildings, often in a circular or polygonal plan. However, many watermills have been successfully converted to a variety of uses including museum, restaurant, gallery or workshop use, even with their machinery intact. Moreover, this may provide an attraction to visitors which may be necessary to make a scheme financially viable.

At the same time there remains a division between conservationists who argue for maintaining both the building and its machinery (Fig. 6.2) and some LPAs who have regarded the visual amenity of the external appearance of such buildings as being the overriding conservation factor. Nevertheless, the annual reports of the SPAB Wind and Watermill Committee indicate a number of examples where such buildings have been 'protected by other means' such as ownership by a BPT. This will often be the only acceptable solution in conservation terms as these buildings tend not to be very flexible to re-use without destroying their historic character both internally and externally.

Industrial buildings

Many large industrial mills and factories, breweries and warehouse buildings of the eighteenth and nineteenth centuries provide considerable opportunities for beneficial re-use (SAVE, 1990a). Virtually every type of modern property occupier has taken space in recycled industrial buildings since the end of the 1970s. They have become open to re-use because of their inherent character, large open floor spans and sufficient strength to accept loading demands (usually much lighter than they were originally designed for); many also stand in waterside locations which has proved an attractive draw.

Fig. 6.2 Warwick Castle Mill, built in the Gothic style in 1768. The corroded cast iron fins of the mill wheel have been replaced by wrought iron and further repair works have been carried out to the masonry walls and mechanical and engineering installations.

At the same time they may stand in derelict or depressed areas and due to redundancy may have suffered from vandalism or have fallen into a severe state of disrepair. A large number of industrial buildings may suffer this fate in a concentrated area. In order to preserve one building it may also be necessary to preserve other nearby buildings before the area is improved sufficiently to attract new users and to make such projects economically viable.

There are also practical problems in converting industrial buildings in conservation terms. Although the external façade, frequently with regular rows of windows conveniently placed on each floor, may not present difficult problems in conversion schemes, it is often the internal construction of industrial buildings which provides the greatest interest. Timber, iron, steel and even concrete used in beams, columns and frames often have a technological as well as an aesthetic value. Thus concealment in order to divide the space or to satisfy fire safety regulations can damage the integrity of a building and cause problems in terms of negotiating LBC.

While maintaining large open spans may be possible, as evidenced in the case of Salts Mill in Saltaire (Fig. 6.3), converted into a gallery to house works by the artist David Hockney, this type of solution will not be available in the majority of cases. Moreover, some deep plan buildings may be inflexible to change due to the difficulty of providing adequate levels of natural light to central areas which may require dividing up. While the creation of inner courtyards or atriums may resolve this problem, such intervention into the historic fabric may not be acceptable.

Agricultural buildings

A Ministry of Agriculture study in 1985 estimated that there were then over 600,000 historic farm buildings, built before 1900, in England and Wales. However, many of these buildings have become redundant because of change in the agricultural industry and no longer earn their keep on the farm. While relatively few of them are completely disused, they are frequently used for low-level storage purposes which does not justify significant expenditure on repair and maintenance (Harvey, 1990). Consequently, the problem of redundancy spirals into one of neglect, placing such buildings at risk.

Yet English Heritage, in a statement on *The Conversion of Historic Farm Buildings,* has stated that the issue of conversion has become one of the most pressing and controversial subjects in historic building conservation (English Heritage, 1990). In particular, the massive increase in the number of proposals to convert redundant buildings into residential accommodation during the 1980s led to calls to halt this form of re-use entirely due to the consequent loss of character or integrity which has often ensued. Moreover, SAVE published a report, entitled *A Future for Farm Buildings,* to highlight the problem and illustrate other uses which have been successfully incorporated in conservation terms (SAVE, 1988).

While residential conversion is often seen as the most economically viable solution to redundancy, it may have a detrimental impact in conservation terms. Residential conversion tends to destroy much of the original fabric by creating new openings and by the removal of structural elements. It frequently results in the disruption of walls and roofs with new doors and windows and breaks up rooflines by the addition of chimney stacks and

Fig. 6.3 Salts Mill, Saltaire. Now houses a print collection of the artist David Hockney; this did not necessitate any significant alterations.

dormer windows. The interior of buildings is changed by the removal of traditional fittings and subdivision to provide rooms. It also leads to the creation of an enclosed domestic plot, removing a building from its agricultural setting and disrupting the integrity of a farm steading.

However, in some instances residential conversion may be possible without major destructive changes, such as where the buildings concerned are two-storeyed and have sufficient openings and internal partitions to permit conversion without significant loss (English Heritage, 1990). Moreover, some commentators have criticised English Heritage's advice to LPAs to maintain 'a general presumption against residential conversion' and its preference for community halls or commercial premises with less potential for design corruption. There are many examples of conservation architects designing schemes which have recognised that the key to conversion is simplicity and maintaining spacial integrity (Morgan, 1991). Yet this desire is frequently more easily achieved by other alternative uses.

The need to maintain the local economy of rural areas as the agricultural industry declines means that alternative business and employment opportunities must be found. The re-use of farm buildings may serve this purpose. Moreover, workshop space, for example, is more likely to be compatible with the historic fabric and may attract funding from the Rural Development Commission. Community uses such as halls, meeting rooms and veterinary surgeries or simple bunkhouses may also allow retention of the open volume of a building. Although these uses may not always be a money-making option they may cover maintenance costs, which is often all that a landowner requires. Farm buildings may also create possibilities for office use in an open plan format although this may be more difficult where the buildings are located some distance from established commercial centres. Nevertheless, the increased use of advanced communications systems now available has made opportunities for 'tele-cottages' where location is less important (Ritchie, 1990).

The letting of redundant farm buildings may have advantages over sale in that farmers will be able to maintain control over their holdings as well as gain a long-term income. This will assist in maintaining the integrity of farm steadings where a number of buildings exist. The Ministry of Agriculture, Fisheries and Food (MAFF) *Farm Diversification Grant Scheme* also provides grants to assist existing agricultural businesses to develop alternative commercial uses which may assist in this purpose.

While a number of different options may be chosen, the initial assessment should also consider the possibility of continued agricultural use for apparently redundant buildings. This is the most desired option in conservation terms but is often unnecessarily dismissed. The MAFF *Farm and Conservation Grant Scheme* was designed to maintain efficient farming systems. Grants are available from this scheme to assist with the cost of repairing or reinstating traditional buildings in agricultural use. Elsewhere similar schemes have been devised through other means. By example, two conservation areas were designated within the Yorkshire Dales National Park in 1989 (Swaledale and Arkengarthdale) and 1994 (Littondale) for the purpose of providing a grant scheme for the repair of traditional stone barns (and walls) to be maintained in agricultural use.

Furthermore, imaginative solutions by some landowners have resolved the problem of traditional buildings being incapable of accepting modern agricultural technology. For example, at the Holkham Estate in Norfolk the problem of some existing buildings being too small for present-day needs has been resolved by adding new buildings in close proximity in sympathetic design so that the two buildings work in conjunction with each other. Moreover, the argument often quoted by farmers that modern tractors with high cabs cannot enter openings in existing buildings can sometimes be resolved by lowering the ground level. The advantage of following this course of action is that buildings which have stood for maybe 100 or 200 years can be preserved in use for many years to come with grant assistance and often at substantially less cost than that involved in the erection of new buildings (Lord Coke, 1990).

Domestic buildings

The domestic dwelling is probably the most adaptable of building types. Depending on size and location, houses may be flexible to re-use as workshops, office and business premises, hotels, schools, libraries, surgeries and health clinics, and for institutional use. However, loading requirements may rule out uses which depend upon machinery or other heavy items. Moreover, high flooring loadings to ensure flexibility in use and avoid overloading may require the original structure to be strengthened or superseded with a consequent loss of original fabric and character. The introduction of steel beams should be scrutinised as there are many ways to provide strength and in many cases all that is necessary is to stiffen floors and weak joints in timber members (Baxter, 1985).

The problem of loading is perhaps greatest with office or business use, often the first choice for re-use in urban centres, which utilises computers and file and document storage rooms. This is exacerbated by investment institution insistence on the highest British Standard loading requirement of 5.0 kN/m^2. Yet in the last 20 years information technology has improved with computer machinery being less heavy and smaller than previously and the ability to store considerable information on small disks. Thus, in practice, a loading requirement of about 1.6 kN/m^2 may be all that is necessary (Redman, 1994). Moreover, recent research findings suggest that while certain isolated areas may require a high loading value, loadings in excess of 2.5 kN/m^2 are not generally necessary in refurbishment work (Hume, 1992). In fact, in many instances office use will only require a loading requirement similar to the original domestic use.

In other respects change should be limited to the requirements of the user or that needed to attract an occupier. The alteration of floor plans, reduction of high floor to ceiling heights, and introduction of modern services are requirements often demanded by institutional property managers to create prime office accommodation. However, as with loading considerations, such changes may be damaging to the character of a building and may not be necessary to create an economically viable unit.

Country houses

Large country houses are flexible to re-use for institutional purposes such as private schools, colleges or care homes which will largely retain the original layout. In some instances they may be used for visitor attractions, but English Heritage and the National Trust are not able to save all buildings of this type that are under threat. Moreover, this course of action often produces a stale feeling caused by the loss of actual inhabitants (Binney and Watson-Smyth, 1991).

Subdivision into apartments for residential use is often the most economical and easy option for the re-use of country houses. However, this choice may pose problems with respect to the internal layout and finishes. A number of schemes have resolved this by providing units through horizontal division, avoiding the need to create new staircases in each unit and using existing corridors for communal use. Alterations which may affect decorative finishes, such as by the lowering of ceilings and the creation of rooms within rooms, may be acceptable if the opportunity to reverse the works, through recording or other means such as safeguarding original plasterwork within voids, is allowed. On the other hand, vertical subdivision may be argued as a better approach to retain the integrity of the internal layout. This may enable the retention of principal rooms and allow units to have attics and cellars as well as using existing external doors for private access. Whichever course of action is chosen, the most successful conversion schemes in conservation terms are those which produce a limited number of apartments that retain, as far as is possible, the original historic quality. This is in contrast to the route chosen by many developers, i.e. to break up the internal layout and finishes in order to provide an enormous quantity of flats (Gillilan, 1994).

School buildings

Many old school buildings have been closed in recent years for various reasons. In rural areas in particular, declining numbers of school children and government policy to save costs have made small school buildings redundant. Many of these have been successfully re-used as libraries or other community uses, including adult education centres, without the need for significant alteration. Conversion to residential accommodation is also common and some buildings may lend themselves to workshop, studio or office use.

Railway buildings

Railway buildings may be flexible to different forms of re-use depending upon their size and location. Buildings within small rural locations often resemble cottages and therefore require little external alteration for residential re-use. Station buildings have been adapted for offices, workshops, restaurants or public houses, although the removal of platforms and track-beds have made the former use unrecognisable in some instances (Johnson, 1988). The increasing number of private railway systems opening up by enthusiasts of the steam engine also now have an important role in preserving the rural railway heritage. Larger railway termini in town centres have also been successfully re-used where the most significant item of interest is often the arched spans of iron, wood and glass roofing. Such spaces may be adaptable for retail use within or for larger space users such as the conversion of Manchester's Central Station into an exhibition centre which retains the spacial integrity of the former station.

Hospital buildings

Redundant historic hospital buildings, particularly from the nineteenth century, are now providing an opportunity for re-use following changes which have taken place in the National Health Service in recent years. Many were commissioned from leading architects and in addition to their individual design qualities some have extensive landscaped grounds and their location on the outskirts of towns provide a suitable location for re-use (SAVE, 1990b). There are opportunities to adapt such buildings to a wide range of uses, including private and housing association residential apartments, private care homes, and business use. There is a temptation to redevelop these sites but government policy has tended to restrict this approach, especially when a building is located within an area designated as green belt land. The main problem with any re-use strategy is to find a use or uses which can fully utilise the large amount of space to be found in such buildings without damaging decorative interiors by careless sub-division.

Public utility buildings

Buildings designed for public utilities tend not to be malleable to re-use. Power stations and gas works are often too large or otherwise do not lend themselves easily to conversion, as is evidenced by the landmark building of

Battersea Power Station which has been considered for a rubbish incineration scheme, but rejected on pollution grounds, and an urban theme park. Yet the privatisation of regional water boards in 1990 has provided a promising opportunity for outdated water pumping houses. Although large in size, with the presence of historic pumping machinery possibly limiting the scope for re-use by being an integral element of interest in such buildings, the mothballing of historic plant *in situ* or removal to a museum site combined with new uses has been found to be successful. Examples of new uses have involved conversion to housing and, in the case of a floodwater pumping station in the London Docklands, to accommodation for musical rehearsals (Pearcy and White, 1989).

Theatres and cinemas

These buildings are most conducive to preservation for their original use. Many of these buildings have been lost or abandoned during the last 50 years. Yet for those that remain many have interiors that are intact despite closure. The Theatres Trust, established in 1982 with statutory recognition requiring notification of any planning applications affecting them, has played a major role in restoring historic theatres which have often survived in almost unaltered condition through temporary storage use or as bingo halls. Many theatres have been re-opened in recent years by specially formed trusts or companies although the theatre business today is financially risky and unpredictable. Re-use for the original function often depends on the availability of arts subsidies (Earl, 1994).

With the upturn in cinema attendance figures in recent years there is the prospect of buildings with historic interiors, the most important being in the Art Deco style of the 1920s and 1930s, being brought back into use. However, the current vogue for small multi-screen cinemas creates pressure for subdivision. Where internal fittings can be retained and new works are reversible, re-use in this manner may be acceptable in conservation terms (Binney and Watson-Smyth, 1991). The only real alternative to preserve both types of buildings is to use them for other live entertainment purposes such as a concert hall.

Feasibility studies, grant applications and VAT

In reality there is no point in considering a new use unless it will produce some financial return. Moreover, for a LPA to consider granting LBC for alterations necessary to allow a new use, it may be necessary to show that sufficient return will be produced to ensure that a building can be properly maintained and preserved in the long term. A bank or other funding body, including grant-aiding organisations, will also need to determine whether a project will be a sound investment for money loaned or otherwise needed to make a project viable. However, in order to obtain necessary consents and grant aid from public sources the scheme proposed must be proven to be the best use in terms of the integrity of the building as well as financial return. Thus a feasibility study of the cost of works and potential return from potential new uses will be required as part of the initial project assessment.

The commissioning of an architect or chartered surveyor, preferably with experience in conservation work, will be necessary to provide a feasibility study. This will require an initial inspection to note structural, repair and other problems, advise on the most suitable form use (if the original use is not to be continued), determine optional alterations in this context, and estimate the consequent end value in light of local market conditions following the completion of works.

Viability studies, as with all valuation and appraisal work, can never be treated as an exact science. The purpose of the study is to assess the parameters of profit or loss of a project according to assumptions made on costs and income. The format of a study will vary according to the particular practitioner who is commissioned. Moreover, the RICS research report *Development Valuation Techniques* has highlighted the different approaches actually used in practice and provides an invaluable guide to techniques used in viability studies (Marshall, 1991). Nevertheless, most studies will follow a similar approach by forecasting the relative expenditure and revenue items which are likely to be incurred.

In cases involving small buildings, such as may be undertaken by a BPT, the appraisal may take the form of a relatively simple budget statement of income and expenditure and an allocation of these items in the form of a cashflow statement according to the timetable of the project (Weir, 1989). Particularly in the case of a BPT it may not be necessary for a project to provide a vast profit, but it should be remembered that any gain can be used to help finance other projects. However, in cases where a major or complex project is proposed, such as for the re-use of a large redundant building, it is almost inevitable that this type of scheme will only be undertaken by a private developer who is willing to take a risk on the large sum of money to be invested. In this situation the scheme proposer will require a profit in return for skills utilised, overheads and the risk involved. Some developers calculate their profit margin by taking a percentage of the capital value of the completed project. But it is more usual to assess this by taking a percentage of the total cost estimate, normally in the range of 12–21 per cent on costs (Marshall, 1991) although a higher percentage may be required from high-risk projects. Moreover, historic properties are frequently regarded as a greater risk compared to other building work, due to the greater incidence of unknown factors which may create problems in a project. For instance, a known outbreak of dry rot within a building may turn out to be more problematical or more extensive once work has commenced; similarly, a find of archaeological significance may be made during the course of works causing delay and additional costs.

The approach that may be followed in connection with a new use proposal for redundant historic buildings is best explained by means of a hypothetical case study.

> **Example** A feasibility study is required to see whether it would be viable to convert a nineteenth-century listed former shipping offices in a riverside location within a conservation area in a provincial city into residential apartments. The building comprises a basement and four upper floors in four wings with a central courtyard. The scheme involves stonework repair and cleaning of external masonry, replacement of windows as original with double glazing, replacement of defective rainwater disposal goods, resolution of ground subsidence affecting one wing, major repairs to the roof and structural works required to remedy an infestation of dry rot partially affecting two wings of the building. The interior timber-pannelled entrance hall and main stairwell and existing corridors are to be retained and repaired but the majority of the interior is to be reconstructed, including the provision of car parking in the basement, four lift shafts and landscaping of the courtyard for communal use.

The first stage is to carry out an initial feasibility study to determine whether the project is likely to show a profit. This is carried out via a residual appraisal of costs and revenue (Fig. 6.4). All figures quoted are arbitrary and should only be regarded as being for illustrative purposes (Fig. 6.4).

The site cost is represented in the appraisal even if the building is already in hand. It is essential that the opportunity cost of the project is identified with respect to other areas of enterprise including other similar schemes (Rees, 1988). Moreover, even if the property is to be retained in ownership the acquisition price or value before the conservation project is commenced should be represented as it may need to be identified for tax purposes. Where a building is not owned it may be necessary to ascertain a realistic purchase price, bearing in mind the relative costs and end value of the proposed project. Thus an appraisal as indicated above may also be utilised to determine the price by identifying the residual sum from the income to be derived after deducting the expenditure incurred as being an amount available for the purchase price plus the incidental costs of purchase including valuation fees, conveyancing costs and stamp duty. The cost of purchase is normally spread over the whole time period of the project.

The appraisal may have to provide for pre-development costs such as planning and building regulation fees and the cost of undertaking a feasibility study. More significantly, owners of historic buildings that have become redundant to their original use may seek to find temporary occupiers in order to secure some income from the property. This can create significant problems in valuing the property and in obtaining vacant possession, thereby delaying the commencement of works and resulting in additional costs, including compensation payable to any existing tenants under Landlord and Tenant legislation (Longden, 1989).

Fig. 6.4 Example of a feasibility study

Costs		
1. *Site purchase*	£	£
Site cost	400,000	
Legal costs (1.5%)	6,000	
TOTAL		406,000
2. *Building costs*		
Repair/new work	1,433,000	
Contingency (10%)	143,300	
TOTAL		1,576,300
3. *Fees*		
Preliminary costs and planning	20,000	
Professional fees on building costs (Architect, Surveyor, Structural Engineer, etc.) (16.5%)	260,090	
TOTAL		280,090
4. *Finance* (16% p.a.)		
Site purchase for 2 years	140,314	
Building cost over 18 months x 1/2	185,606	
Professional fees for 2 years x 1/2	44,814	
Voids and insurance, say	50,000	
TOTAL		420,733
5 *Sale costs*		
Agents fees (2.5%)	81,600	
Legal fees (1%)	32,640	
Advertising	50,000	
TOTAL		164,240
SUB TOTAL		2,847,363
6. *Developer's profit*		
On costs at 20%		569,473
TOTAL COSTS		3,416,836
Revenue		
Sale of completed apartments		
Each wing comprising 8 x 2 bed. apartments and 3 x 3 bed. apartments.		
	£	£
32 x 2 bed. at £72,000	2,304,000	
12 x 3 bed. at £80,000	960,000	3,264,000
Grant Aid	150,000	
TOTAL REVENUE		3,414,000
Loss/Profit		
3,414,000 − 3,416,836		− £2,836

The most critical factor is the cost of works. As the works will be considerably different to those concerned with a standardised new building type an estimate of construction costs should normally be provided by a quantity surveyor. Conservation work tends to be more specialised and expensive compared to new work and it is here where the greatest risk is incurred. Thus a relatively high contingency element is provided for, which is likely to be in the region of 10 per cent (Cantlie, 1990) or even higher with a particularly complex and difficult building conservation project (Davey, 1992). Although the profit margin makes some allowance for risk it is often best to be cautious when dealing with historic buildings due to the possibility of miscalculating the extent of structural problems and conservation remedies required. Moreover, in some instances the opening up of the fabric may reveal hidden older work which may require a new strategy to be adopted. Any delay in this context may add to costs, particularly if the method of procurement has not fully anticipated this possibility. It is also usual to limit the cost of building works to the time period in which they are expected to occur, rather than for the whole period of the project, and to reduce this by a fraction of one-half, two-thirds or three-quarters to reflect the fact that payment for work will be made in stages.

Professional fees will vary considerably depending on the type of scheme and the problems likely to be encountered. The percentage of costs taken for fees will normally be in the range of 12–17 per cent. With historic buildings it is more likely that specialist advice will be required from professionals who have experience in dealing with such buildings and their inherent problems, and therefore the fees are likely to be in the higher part of the range indicated. For instance, advice from an appropriate structural engineer may be required to deal with a particular problem of alteration where the historic fabric may be threatened by a certain course of action. In particular, the question of floor loadings from a proposed new use may require an imaginative solution if the character of the building is not to be damaged.

In other respects the type of building and the proposed use will determine the range of professional advice required. In addition to an architect and surveyors to deal with design, planning, valuation and cost issues it may be necessary to account for fees for structural engineering and mechanical and electrical engineering advice, project management, fund supervision and other specialist advice. As with the financing of the building works the interest on money borrowed for fee payments for professional services should be reflected over a fraction of the whole project time period to account for fees being paid in stages.

Sales may take longer than initially anticipated and once the building contractors have completed works there will be a responsibility to insure and heat unsold apartments, which may incur additional costs.

Sale fees include estate agent fees and legal costs. These will be in the range of 1.5–4 per cent. Where the end use is one which will result in the building being leased, for instance in the case of office use, a letting fee of around 15 per cent of the initial rent should normally be taken into account in the appraisal. The level of estate agent's fees will depend on whether one or more than one agent is instructed to find purchasers or tenants.

The revenue element of the initial appraisal is more usually considered before costs, but due to the shortfall in the imaginary scheme it has been represented at the end where grant aid against repair costs has been included to ensure a negligible deficit on profit. Although a residential end use, as indicated in this example, would result in the sale of long leases rather than a freehold interest, the ground lease income from the individual apartments has not been included in the appraisal as the amount will be negligible bearing in mind the need for a management agreement to deal with costs associated with common parts and rent collection. Where the end use does not envisage a sale – for example, where the premises are to be let to business occupiers – the capital value of the finished scheme would be ascertained by accepted property valuation techniqes via the capitalisation of rental value (Issac and Steley, 1991).

The next stage is to break down the costs into a cash flow over the whole period of the project. For large projects these are relatively easy to produce by feeding the refined figures into one of the now many computer spreadsheet programs that have been developed for financial appraisals and viability studies. By this method the figures are generally represented on a horizontal basis (Fig. 6.5), with the provision of a running total. A simple cash flow may be used which averages out the costs over monthly or quarterly time periods. Alternatively, an attempt may be made to pinpoint the time when costs will actually be incurred. The advantage of the cash flow approach over the initial residual appraisal is that it indicates period payments, including interest charges, which will be an essential requirement of lending institutions and other funding bodies. A more sophisticated means of cash flow appraisal may be utilised by discounting the balance of all the costs and income as outflows and inflows at a target rate, usually at the cost of finance, to find the opportunity cost of different projects (Marshall, 1991).

The interest on finance borrowed, quoted as an annual rate, is charged quarterly. Thus, an annual borrowing rate of 16 per cent is charged at 3.782 per cent per quarter on costs divided into periods when they are likely to be incurred. In practice a longer start-up period may be required, particularly where existing tenants are to be relocated. Professional fees may be high at the commencement of the project to account for the feasibility study itself and for a substantial part of the design fees, which may be payable upon completion of the relevant drawings and plans.

The cash flow provides for the sale of 12 × 2 bed and 4 × 3 bed apartments in the seventh quarter, and the sale of the remainder in the last quarter. Thus the cash flow provides a higher profit than the initial appraisal. Yet, in practice, sales may take longer and a delay of six months in selling half the apartments would erode the profit margin, as has been accounted for in the initial appraisal. Moreover, it must be reiterated that the example provided is fairly crude and the figures are for illustration purposes only. Nevertheless, an essential consideration in undertaking feasibility studies is to scrutinise how the variation of the estimated figures will affect the

Fig. 6.5 Example cash flow

Annual borrowing rate	16%							
Payments period p.a.	4							
Period borrowing rate	0.037802							
Period number	1	2	3	4	5	6	7	8
Expenditure								
Purchase	406,000							
Demolition	50,000							
Scaffolding	40,000							
Roof works		45,000	135,000					
Internal works			20,000	40,000				
Timber treatment		50,000						
Foundation works		25,000						
Masonry works		30,000	85,000					
Rainwater goods			48,000					
Windows				90,000				
Basements works				50,000	30,000			
Internal fitting					120,000	180,000	60,000	
Advertising				10,000	10,000	10,000	10,000	10,000
Lifts/stairwell					100,000	100,000		
Services					50,000	80,000		
Landscaping							25,000	
Contingency	90,000	5,000	28,000	18,000	30,000	36,000	8,500	
Fees	100,000	30,000	20,000	30,000	30,000	30,000	20,000	10,000
Sub total	605,000	195,000	336,000	238,000	370,000	436,000	123,000	20,000
Balance brought forward		627,870	828,031	1,156,144	1,420,900	1,832,655	2,328,470	1,358,908
Total quarterly expenditure	60,000	822,870	1,164,031	1,394,144	1,790,900	2,268,655	2,451,970	1,378,908

Fig. 6.5 *continued*

Period number	1	2	3	4	5	6	7	8
Income								
Grant aid		25,000	50,000	25,000	25,000	25,000		
Sales							1,184,000	2,080,000
Less agents fees							41,440	72,800
Total income		25,000	50,000	25,000	25,000	25,000	1,142,560	2,007,200
Outstanding balance	−605,000	−797,870	−1,114,031	−1,369,144	−1,765,900	−2,243,655	−1,309,410	628,292
Interest on balance	−22,870	−30,161	−42,113	−51,756	−66,755	−84,815	−49,498	
Cumulative balance c.f.	−627,870	−828,031	−1,156,144	−1,420,900	−1,832,655	−2,328,470	−1,358,908	
Balance developer's profit	£628,292							

profit/loss margin. A sensitivity analysis should be carried out to evaluate the critical variables within a project. These include the building costs, interest charges, yields (for the capitalisation of rental values), sale price or end value. The most vital factor for scrutiny in the case of historic buildings is likely to be the building costs and is the principal reason for a high contingency element. By using an appropriate computer program a sensitivity analysis can widely assess the critical variables. At minimum break-even, down-turn and up-turn forecasts should be provided for the lending institution and any grant-funding body.

The sensitivity analysis may indicate the need for additional funding which may require investigation of the various sources of grant aid (see Chapter 4). In consideration of making an application for grant aid, it is important to be aware that financial support is finite and that other schemes will be competing for limited resources. The onus is on the applicant or relevant professional advisers to provide all necessary information. This should include detailed and realistic estimates of the likely cost of eligible works which, in the case of conservation grant awards, will usually be limited to repair works rather than alterations or improvement work. An apportioned amount of preliminary costs and professional fees may also be considered in the grant application appraisal. Priced schedules of work or bills of quantities, depending upon the size and complexity of the project, should be submitted together with photographs, plans, drawings and specifications to show the full extent of works (Davey, 1992). In practice, an initial feasibility study and cash flow appraisal may differentiate between relevant work and non-eligible work and, in any event, it may be necessary to apportion repair work for VAT purposes. In the case of grant applications made by BPTs, it may be necessary for the feasibility study to consider costs in relation to an existing rolling programme of projects as capital may be released from other completed projects.

VAT costs should also be considered in the feasibility study, but due to the complex rules regarding zero-rated works in the case of listed buildings they have not been indicated in the initial appraisal and cash flow example. The position in relation to VAT must be considered according to Customs and Excise leaflet 708/1/90, *Protected Buildings (Listed Buildings and Scheduled Monuments)* and the provisions of Group 8a of Schedule 9 to the VAT Consolidation Act 1994. Specific VAT relief for *protected buildings* may occur in two situations, i.e. in relation to the provision of construction services (where a building owner obtains the services of a building contractor to carry out work) and in relation to property transactions (where an owner carries out the work and then sells the freehold interest or makes a sale via a long leasehold of greater than 21 years, as may be the case where a developer undertakes the works). A *protected building* is one which is designed to remain as, or become, a dwelling or number of dwellings or is intended for use only for a *relevant residential purpose* or a *relevant charitable purpose*.

In the first case the supply of goods and services in connection with an *approved alteration* to a *protected building*, other than the services of an architect, surveyor, or any other person acting as a consultant or in a supervisory capacity, are zero rated. An *approved alteration* includes works of alteration which cannot be carried out unless authorised by LBC (or Scheduled Monument Consent). Other alteration or repair work does not benefit from this provision except where an element of the works requires consent, in which case an apportionment is required to determine which works may be zero rated. Thus, where a three-storey dilapidated listed house is to be converted into a ground-floor shop with two flats above via the granting of LBC, the alteration elements of the works, which have required LBC and are associated with the flats only, can be apportioned from the repair element of the works to the flats for the benefit of VAT zero rating with respect to the contractor's costs. The part of the property to be used for retail purposes does not qualify for relief as the use is not a qualifying use.

The term *relevant residential purpose* includes: a residential home or institution for children; a residential home or institution providing care to persons by reason of age, disablement, dependence on alcohol, drugs, or mental disorder; a hospice; residential accommodation for students, schoolchildren, or for members of the armed forces; a monastery, nunnery, or other similar establishment; and certain other institutions which are the sole or main residence of at least 90 per cent of their residents. The term *relevant charitable purpose* includes use by a charity other than for business purpose where supplies of goods and services are made for a consideration whether or not for profit (for example, a BPT is *not* a relevant charitable purpose), or a village hall or other similar social and recreational facilities for a local community.

The second case concerns the zero rating of taxable supplies of interests in land or property where there is a grant of a major interest in a *substantially reconstructed* relevant qualifying building, i.e. a *protected building*, by the person who substantially reconstructs it. There are two tests of *substantial reconstruction*. First, the 'Demolition' test, which is where the reconstruction incorporates no more of the building – as it existed before the works commenced – than the external walls, together with any other external features of architectural or historic interest. The second test is known as the '60 per cent' test, which requires that either the building is reconstructed on its facts or that at least three-fifths of the cost of works are related to *approved alterations* if carried out by a VAT-registered person. The second case, a VAT zero-rating situation, allows full relief of all input tax incurred on a project including VAT charged on the fees of professional services. An apportionment will be required where the use of the building is only partially qualifying for relief, although it must be established first that the 60 per cent rule applies to the works on the whole building.

There is also potential for recovery of VAT in the case of non-qualifying buildings as it is possible to make an *option to tax*, i.e. to charge VAT on the sale or letting of non-domestic buildings so that the tax incurred may be recovered by the person making the transaction. However, if the purchaser or lessee is not VAT registered this person will not be able to reclaim the VAT charged on the transaction. As the *option to tax* is irrevocable, this course of action must not be taken lightly as it may create a disadvantage to partially exempt or non-VAT-registered potential tenants or purchasers as compared to other buildings in the property market, and therefore reduce the attractiveness of a building.

Clearly the potential liability to VAT has significance to the viability of a conservation project, both in terms of the works actually undertaken and if the building is to be subject of a property transaction after the works have been completed. The complex rules associated with VAT legislation, which have only been explained in brief terms, will make it necessary to seek expert advice which in turn can be considered in the cost analysis and planning of a particular project. Projects involving mixed uses will require scrutiny in terms of partial exemption and apportionment, and in other respects recovery of input tax may be achievable according to certain *de minimus* rules.

Formalising and managing a project

In order to commence a conservation project various tasks have to be carried out. The preparatory works of the initial project assessment have been indicated. If a property is not in-hand, the legal process of purchase has to be sorted out and building insurance arranged for the building itself and for site accidents before and after the contractor's liability period. A project manager and team have to be appointed and a project brief devised. Relevant official approvals must be sought and tendering and contractual arrangements entered into. After the choice of contractor(s) has been decided the works have to be supervised and financially monitored.

The project manager and team

The skills required for a conservation project will largely depend on issues of size and complexity: whether it relates to small-scale repair work, a large repair or restoration scheme and/or the accommodation of a new use. Small-scale repair or alteration work may only require a builder or other individual with building craft skills, and possibly a building surveyor or architect to oversee the works and deal with matters of design. As the works become more complex or large in scale a greater amount of control and professional skills are required for design and administration matters and supervision of contracted or subcontracted works. Furthermore, the range of skills may have to be varied depending on the nature and type of building involved.

Invariably a coordinator or project manager will be required to oversee the project from inception to completion. The project manager is usually given authority to act as the client's sole representative with responsibility to coordinate the relationship between the client, project team members, contractor and others involved in the project.

The project manager's role in conjunction with the client may involve a variety of tasks. These include arranging the appointment of consultants and team members, their instructions, establishing reporting procedures, chairing progress meetings and attending site meetings as necessary; assessing contract and associated funding, and viability criteria; organising preliminary works on site acquisition, project planning, and preparation of the initial project design brief; financing the project; dealing with on-going financial and legal matters; liaising with the contractor throughout the works period on timing, quality and costs; controlling additional contractor requirements and providing resulting instructions to team members; liaising with agents and solicitors on marketing, sale and letting issues for future occupation.

The project manager's role associated with the project team may involve the coordination of various activities. These include the preparation of the project brief and applications for official approvals; critical appraisal and monitoring of the project at all stages as defined in a master plan of work; financial planning and cost control; selection of the most appropriate form of building contract and appointment of building contractors; and overseeing the progress of the contract, including the control of variations, monitoring reports made by the team members and progress on site, and monitoring the availability of materials, labour and services.

The choice of professional to act as project manager will be dependent on the relative skills required. For small-scale projects such as may be contemplated by a BPT, a conservation architect may oversee the project. However, chartered surveyors are increasingly undertaking this task, particularly in complex schemes which are complicated and require firm cost control. A building surveyor may be an appropriate professional to undertake the role on the basis of design and conservative repair knowledge. Alternatively, complex projects are often managed by a quantity surveyor in order to ensure efficient financial control, particularly where the relevant professional has significant historic building work experience. Moreover, all professionals learn the process of project management through experience, and relevant experience with historic buildings is a vital factor in conservation work. A good project manager is one who is prepared to delegate but also respect the views and experience of other members of the project team.

Major projects may also be managed by a specialist development or building management company, via a management fee contract, which benefits from having an in-house team who work together with confidence derived from shared experience. However, selection of a company should be based on knowledge of its previous conservation work, and if this option is chosen the client may be best served by retaining an independent quantity surveyor to advise on financial control matters and conditions of contract including the need for variations, and a design consultant to advise on repair or alteration solutions (Davey, 1992). The benefits of shared experience may also be achieved by bringing together members of the project team who have previously worked together on conservation projects.

Whether a project will have a designated project manager as well as a design team leader will be dependent on the scale of operations. In small projects the team leader may act as manager. The team leader will have responsibility for planning the works (in conjunction with anyone in a management role) and more specially for the management of the project team and coordinating applications for official consents. The team leader may be a conservation architect, but building surveyors, who are directly trained in scrutinising building problems rather than creative design matters, are now being regarded as the most appropriate professional to undertake this task.

The other building professionals in the team are likely to include a quantity surveyor, a structural or civil engineer, mechanical and electrical engineers, and a clerk of works or resident site representative. Some professionals may act in a consultancy capacity, particularly for engineering matters or concerning planning, valuation or management matters. Advice may also be required on public health matters such as in relation to sanitary installations. In some instances the team may include a landscape architect, such as where the project involves a historic country house with landscaped grounds; a conservationist, where internal decorative schemes or works of art are to be restored; an archaeologist, to understand the development of a historic building, including the methods and materials used in its construction and to advise where archaeological remains are known to be evident in the site or are found during the course of works; a historian and an industrial archaeologist, to advise on historical events in relation to a building, and its development and use over time, including the use of machinery or processes associated with a building.

The project brief

A brief for a particular building cannot be prepared until the initial project assessment has been carried out. Thus decisions must be taken, bearing in mind the research carried out in relation to the building's history, location, condition, flexibility to use and viability considerations in order to devise a philosophically defendable brief. Solutions considered may raise practical, economic and philosophical problems and a balance must be achieved between these before the brief is written in its final form. The advice of conservation and planning officials – and, where necessary, English Heritage officials, other conservation experts and grant-aiding bodies – should be sought in addition to the views of the project team members before the brief is finalised.

The brief will provide a framework for the project activities and a basis for instructing a contractor on the extent of works and how they are to be performed. It should summarise the main features of the project including the building's history, special character, condition, repair and structural works required, essential features to be conserved, the extent of alterations, the materials to be used and the proposed end-use (if applicable) (Weir, 1989). It should further prioritise work through a plan of work in a logical sequence from 'immediate' and 'urgent' needs for safety reasons and for the prevention of active deterioration, to 'necessary' and 'desirable' needs appropriate for the building in its present or proposed use (Feilden, 1994). From the brief a *specification* of materials and works to be carried out in a project may be written. This statement can be provided via a standard specification such as the National Building Specification (Davey, 1992). The brief may also identify the roles of consultants and team members.

Contractual matters

The choice of contractor to be employed will depend on a variety of factors. Small repair projects may only require individual specialist skills. Large rehabilitation projects may require a main contracting organisation and domestic or nominated subcontractors. Whatever the scale of operations an assessment of potential contractor's reliability, financial standing and suitability to undertake conservation work must be made.

The members of the project team or other local professionals may have experience of contractors although they may be wary to recommend a particular contractor if the experience of successful work is only slight or limited to one commendable project. Nevertheless, the evidence of awards such as from the Civic Trust or the RICS *Building Conservation* and *Craftsmanship in Building Conservation* Awards may indicate good examples of works undertaken by contractors. The local conservation officer or planning department and may also be familiar with contractors that have had a good track record in conservation work, and although they will not be able to recommend named firms it may be possible to obtain a list of contracting firms which can subsequently be investigated. Similarly, local and national conservation bodies and local branches of building professional bodies may be consulted for names of possible suitable contractors.

Advice may also be sought from various advisory organisations regarding affiliated firms that have been vetted according to specified standards. For overall operations these may include the *Federation of Master Builders* and *The Guild of Master Craftsmen.* For individual skills these may include such organisations as the *British Wood Preserving and Damp-proofing Association, The Glass and Glazing Federation, The Guild of Architectural Ironmongers,* the *Lead Sheet Association,* the *National Council of Master Thatchers,* the *Master Carvers Association* and the *Stone Federation of Great Britain.* At the same time relatively scarce conservation skills should be deployed where they are most needed. Thus wastage should be avoided where the aspects of the work do not warrant highly skilled craftsmanship while patience may be required to obtain the desired skills in some instances (see Appendix 4).

For tendering purposes sufficient information must be made avaliable by the project brief and specification or other means for contractors to price the works such as a bill of quantities. It is normal for a contractor to be chosen through the process of competitive tendering and this will be an essential requirement where a grant is to be obtained from a public funding source such as English Heritage. Small-scale repair projects may only require three or more specialist firms to be asked to tender. As the works become more complex and costly the number of firms invited to tender should be increased to ensure a competitive edge.

One of the main problems associated with employing contractors to work on historic buildings is the question of unknown factors in terms of the extent of necessary repairs, or other factors which may delay or add to the cost of the works such as a significant archaeological discovery being made in the process of the works being carried out. For this reason, and to save costs, it may prove beneficial to stage the work in two contractual periods. The initial period may be used to provide temporary support and protection and to provide access to concealed areas so that a comprehensive schedule of repair works can be written. This should provide a clearer basis for letting the main works of repair and alteration. It is also usually necessary for a contract to account for the need to vary or modify the works which may arise during the course of the contract, for example, as may be required by an English Heritage inspector when the progress of grant-aided work is examined. Further consideration is required in determining the most appropriate form of contract for a conservation project. The principal forms of contract which may be considered are *the prime cost* contract and those based on a *lump sum.*

A *prime cost* contract (also known as *cost plus*) is one which reimburses a contractor for the actual cost of materials, plant and labour with an additional sum being provided as a lump sum on costs or as a percentage of costs to cover overheads and the contractor's required level of profit. In the first instance this form of contract may be considered attractive in the context of works to be carried out to historic buildings as it can easily deal with the unknown factors which may emerge during the course of the works. However, it provides little incentive for the works to be carried out efficiently. Furthermore, unless the works are relatively minor, such as emergency repair work, it is difficult to predict the total costs prior to tender acceptance.

The lack of financial control often results in the *prime cost* contract being rejected as a suitable form of contract for works to listed buildings. Moreover, English Heritage does not generally accept this form of contract which is an important consideration when an application for grant aid is contemplated (Smith, 1990). Yet the risk associated with *prime cost* contracts can be reduced if the contractual conditions permit the appointment of nominated subcontractors for certain specialist works which can be made subject to competitive tendering provided that sufficient investigation has been carried out to prepare the relevant tender documentation and that such work will not interrupt the continuity of the main contractor's activities.

Lump sum contracts offer a greater degree of financial control, allow for competitive tendering and provide scope for flexibility to deal with the uncertainties of conservation work. For small and medium-sized projects a project may be let on this basis *without a bill of quantities*, relying on a detailed specification to provide the information for tenderers to determine the extent of works for pricing purposes. However, this has the disadvantage of the contractor having to carry out additional work in estimating and pricing quantities at the tender stage. There is also the possibility of disputes over the valuation of variations. The risk of disagreement in this context may be reduced by providing a schedule of works, to be priced by tenderers, with the specification.

For large or complex projects where the works cannot be clearly defined at the tender stage, a *lump sum* contract *with a bill of quantities* may be used. This provides a firm basis for valuing variations and for analysing costs for grant applications. Where the precise extent of works cannot be determined before commencement on site an *approximate bill of quantities* may be used. In this instance the contractor will be paid on the basis of work actually performed on predetermined rates, the final of cost of which may not be determined until the works are remeasured and valued. This provides considerable flexibility as is found in *prime cost* contracts but has the advantage of greater financial control by providing a proper basis for competitive tendering and a contractual mechanism for pricing works.

Other forms of contract may be available for conservation work in specific situations such as a *measured term contract*, which may be suitable for works required by large organisations on a number of sites in close proximity over a period of time, or a *management fee contract*, as mentioned previously, for projects of a high cost value. Further detailed information concerning the complexities of contractual matters may be sought from specialist publications with *Building Conservation Contracts and Grant Aid: A Practical Guide* (Davey, 1992) providing authoritative information on conservation work.

Financial control and supervision

Once a contractor has been appointed a date for work to commence on site can be agreed. The person appointed to supervise the works, usually an architect or building surveyor, should make regular checks with the clerk of works or resident representative on progress, including any delays or stoppages due to weather conditions or problems associated with suppliers, labour, plant, and should inspect the work itself. The project manager should be immediately informed of any unsuspected problems so that the matter can be discussed with the project team and any consultants and a decision made swiftly if required.

The person appointed to manage the project will require to be informed of every aspect of the project at appropriate stages, particularly regarding finance so that funding organisations can be informed and requests for payments to the contractor can be dealt with promptly and efficiently. This will require the supervisor's instructions to be sent to the cost adviser, so that variations and adjustments can be agreed and interim payments paid. The cash flow statement will require adjustment as the works proceed and the client will also require regular statements on cost in order to obtain an estimate that is as accurate as possible of the final cost of the project.

After care, management and marketing

A progressive record of the works carried out should have been made, which may be added to the information already known about the building, and retained for future reference as may be required. If the building is to be sold a marketing exercise must have been commenced before the completion of the works. It may also be necessary to devise suitable methods to ensure the future protection of the building through legal covenants, a management agreement and insurance. If the building is to be retained a management plan must be organised for maintenance purposes.

Recording

The *Manual on Research and Recording Historic Buildings* recommends the value of building up a cumulative record of historic buildings as a reference document for historians, conservationists, owners and their professional advisers (ICOMOS UK, 1990). The record of the history and structural condition of the building before the project was commenced should be supplemented by comprehensive details of the works carried out to provide a management tool which may be of assistance in future maintenance work or to provide information for consideration in future projects concerning alterations or a new use. Recording may also be a condition of LBC or in the award of grant funding or in relation to planning permission in affecting matters of archaeological importance.

Any work involving total or partial demolition; destruction (or concealment) of evidence of a building's origin, plant, machinery, fittings or historical decorative scheme; major repairs; remedial treatment over time; and the location of services, should be recorded. New work should be recorded with sufficient detail to explain the reasons for the work, the date of the works and how they were carried out. The provision of a management document or depository for a cumulative record may be supplemented by clear details of subsequent routine maintenance work.

It is important that the record is then preserved in a safe place either with the owner, or with the architect or surveyor who has supervised the works, or in a public repository such as the National Building Record held by the RCHME, or the County Records Office. A record made in the course of a contract may also be sent to the British Architectural Library of the RIBA.

Marketing

When marketing a building for sale following a conservation project, care should be taken to describe the building accurately due to the provisions of the Property Misdescriptions Act 1991. This will require advertisement information to be checked for historical accuracy and in relation to details concerning the building if it is listed, located within a conservation area or if it is protected under the provisions of the Ancient Monuments and Archaeological Areas Act 1979.

The project manager should ensure that proper steps are made to advertise a building to be sold or let by devising a suitable marketing strategy and decide whether this should commence before the project has been completed. In addition to general marketing methods through estate agents and newspaper advertisements, specialist journals may be used to bring attention to the prestige aspects of a particularly interesting historical building. Where a building has been rehabilitated as part of a local 'buildings at risk' initiative the relevant local authority departments dealing with economic development or planning issues may be able to assist in finding new occupiers. Local newspapers may be used to bring attention to a building conservation project, as an issue of local interest, which may indirectly help to advertise a building. Publicity offered by the conservation movement can also help to find occupiers for a completed building conservation project. For example, in addition to the quarterly list of 'buildings of historical interest in need of repair and for sale' issued by SPAB, a list is also provided by the APT of properties complete or nearing completion following works instituted by a BPT.

The importance of effective marketing as an essential part of a conservation project must be emphasised as the key to long-term preservation is to keep a building in use so that there is sufficient reason and financial incentive to maintain it in the future.

Management strategy

Most building contracts include a defects liability period, usually of six months, although this may be varied to suit the particular building or to rectify any defects in the work which have become manifest once the work has been completed and the building has become occupied. After this period the responsibility for maintaining the property lies with the owner or occupying tenant. A planned maintenance and repair programme should be organised with the building record updated in accordance with works carried out in the course of time. It is important that any management document has a defined philosophical approach so that acceptable conservation techniques are utilised in future work.

A building may otherwise be protected from damaging alterations by retaining the freehold and selling a long leasehold or otherwise letting the building, enabling lease terms to control activities for the benefit of the building. Where the freehold interest is sold outright, restrictive covenants can be imposed on the first purchaser and may, in some circumstances, be enforceable against future owners. Alternatively, each successive purchaser can be required to enter into a Deed of Covenant incorporating both positive and restrictive provisions (Weir, 1989).

Where the circumstances do not permit a full-scale repair/alteration project, and a building is at risk through disrepair and vacancy, a good management approach may still be used by carrying out emergency repairs and 'mothballing' a building until such time as a full project is logistically possible. Proactive maintenance can be organised at minimum cost to check moisture ingress by temporary roof drainage systems, unblocking existing systems, or removing potential moisture reservoirs such as wet furnishings and by providing ventilation by unblocking chimneys and fixing windows slightly open or perforating window coverings (Hutton and Lloyd, 1993). Grant aid may also be available to carry out urgent repairs and temporary protection works (Michell, 1988).

As well as carrying out regular maintenance inspections and repair work as required, a management policy should be instituted regarding the safety of the building, particularly in relation to the prevention of fire or other damage (Pickard, 1994). A strategy must be devised not only for the safety of occupants but also the building, fittings, finishes and any items of value including items that have been designed in relation to a historical decorative scheme, such as works of art or antique furniture. Insurance also plays an important role in safeguarding the future of a building.

Insurance

The most acute risk period for a historic building is likely to be during the course of a conservation project. The Windsor Castle fire disaster highlighted the dangers of appliances that produce a naked flame. Losses caused by the inefficiency of contractors must be covered by their own insurance. Nevertheless, the terms of an insurance policy maintained by the owner should be checked when additional risks are likely to arise. Moreover, all factors which may affect premiums should be scrutinised before deciding on the extent and terms of insurance cover. These issues were higlighted in a joint Engish Heritage/RICS statement published in March 1994, which developed from an information sheet devised by the RICS Building

Conservation Skills Panel entitled *Approach to Property Insurance and Insurance Valuations for Historic Buildings* (Sutch and Davey, 1993).

Different levels of insurance are generally available for property. The main problem associated with insuring historic buildings is to determine the appropriate level and type of insurance cover. Although the extent of damage to a listed building may provide a viable argument for granting LBC for full or partial demolition, it must be remembered that listed buildings cannot automatically be demolished without consent. Thus the insurance cover for a particular building may have a bearing on the extent to which a building can be restored following damage.

An *Indemnity Cover* based on the loss of value to an owner is unlikely to provide adequate cover for total loss where the owner opts to insure the cost of a modern replacement building which will be inherently less expensive than the choice to rebuild an old building on a like-for-like basis. Insurance on a *Total Reinstatement* basis for total loss should provide the owner with enough cover to completely rebuild in the same style and quality but will have the highest insurance premium. A less expensive option would be to insure on a *First Loss* basis according to an assessment of the maximum amount of damage likely to be incurred in a single event with the possibility of retaining an *agreed value* for irreplaceable works of art.

No common rules can be determined for all protected buildings as they vary in size, materials, occupation and relative risk. In general terms, an owner is perhaps more likely to decide on a certain level of insurance cover according to financial self-interest rather than by the historic qualities of a particular building. However, it is also important to bear in mind that freedom of action following damage occurring to a protected building may be limited. At the same time English Heritage has recognised that full reinstatement is unlikely to be desirable where over half the historic fabric has been destroyed. Yet it is good management practice to ensure that a building is adequately insured and that action is taken to reduce the relative risks of damage from fire, smoke, theft, vandalism, natural catastrophe or other factors (English Heritage/RICS, 1994).

References

Ashurst, J. and Ashurst, N. (1988) *Practical Building Conservation*: Vol. 1, *Stone Masonry*; Vol. 2, *Brick, Terracotta and Earth*; Vol. 3, *Mortars, Plasters and Renders*; Vol. 4, *Metals*; Vol. 5, *Wood, Glass and Resins*, Gower Technical Press.

Baxter, A. (1985) Sympathetic engineering, *Architects Journal*, Special Issue, 21 and 28 August, pp. 89–90.

Bowles, R. (1992) Development of building structures and principles of repair. Paper presented at SPAB course in *The Repair of Old Buildings*, 28 September–3 October.

Binney, M. and Watson-Smyth, M. (1991) *The SAVE Britain's Heritage Action Guide*, Collins & Brown.

Brand, V. (1992) *Buildings at Risk: A Sample Survey*, English Heritage.

Cantlie, H. (1990) A guide to financial feasibility studies. Paper presented at a conference entitled *The Conversion Of Listed Buildings*, organised by the Historic Houses Association, held at the Royal Institution, London, 21 March.
CIRIA (1986) *Structural renovation of traditional buildings*, Report III, Construction Industry Research and Information Association.
Davey, K. (1992) *Building Conservation Contracts and Grant Aid: A Practical Guide*, E. & F.N. Spon.
Earl, J. (1994) Preserving theatres. Paper presented at a conference entitled *Listed Buildings: Economic and Financial Consequences*, held at the University of Cambridge, 23 May.
English Heritage (1990) *The Conversion of Historic Farm Buildings: An English Heritage Statement* (revised 1993).
English Heritage/RICS (1994) *Insuring your Historic Buildings: Houses and Commercial Buildings*.
Feilden, B.M. (1994) *Conservation of Historic Buildings*, rev. edn, Butterworth Architecture, pp. 183–307.
Gillilan, L. (1994) Take your seat, *The Guardian Weekend*, 27 August, p. 37.
Harvey, N. (1990) Old farm buildings – the background. Paper presented at a conference organised by the Historic Farm Buildings Group in association with the Rural Development Commission, RICS and English Heritage entitled *Old Farm Buildings in a New Countryside*, held at the RICS Westminster Centre, London, 29 November.
Hunt, A. (1989) Electrical installations in old buildings, *SPAB Technical Pamphlet* 9, SPAB.
Hume, I. (1992) Floor loadings and historic buildings, *Conservation Bulletin*, Issue 18, October, pp. 1–2. (See also *Office Floor Loadings in Historic Buildings*, Information Leaflet published by English Heritage, June 1994.)
Hutton, T. and Lloyd, H. (1993) 'Mothballing' buildings: proactive maintenance and conservation on a reduced budget, *Structural Survey*, Vol. 11, No. 4, Spring, pp. 335–42.
ICOMOS UK (1990) *Manual on Research and Recording Historic Buildings*.
Issac, D. and Steley, T. (1991) *Property Valuation Techniques*, Macmillan.
Johnson, A. (1988) *Converting Old Buildings*, David & Charles.
Longden, J.D. (1989) A developer's rationale or conservation – is it investment or art? Paper presented at a conference organised by the RICS Yorkshire Branch Building Surveyors Division entitled *Building Conservation*, held at St Williams College, York, 11 May.
Lord Coke (1990) Old farm buildings – a headache for the landowner? Paper presented at a conference organised by the Historic Farm Buildings Group in association with the Rural Development Commision, RICS and English Heritage entitled *Old Farm Buildings in a New Countryside*, held at the RICS Westminster Centre, London, 29 November.
Marshall, P.J.L. (1991) *Development Valuation Techniques*, RICS.
Michell, E. (1988) *Emergency Repairs for Historic Buildings*, English Heritage/ Butterworth Architecture.
Morgan, A. (1991) Enlightenment in an old barn, *The Independent*, 21 September.
Pearcy, O. and White, P. (1989) The water heritage, *Conservation Bulletin*, Issue 8, June, pp. 4–6.
Pickard, R.D. (1993) Fire safety and protection in historic buildings in England and Ireland – Part 1, *Structural Survey*, Vol. 12, No. 2, 1993/94, pp. 27–31.
Pickard, R.D. (1994) Fire safety and protection in historic buildings in England and Ireland – Part 2, *Structural Survey*, Vol. 12, No. 3, 1993/94, pp. 8–10.
Preston, J. (*et al.*) (1993) Proceedings of the *Secret Services* conference organised by the Association of Conservation Offices (East of England Branch) held on 27 November at Robinson College, Cambridge.
Redman, C. (1994) Consequences for investment. Paper presented at a conference

entitled *Listed Buildings: Economic and Financial Consequences*, held at the University of Cambridge, 23 May.
Rees, W.H. (ed.) (1988) *Valuation: Principles into Practice*, 3rd edn, Estates Gazette Limited, pp. 387–414.
RICS/English Heritage/IPD (1993) *The Investment Performance of Listed Buildings*.
Ritchie, M. (1990) New industries in old farm buildings. Paper presented at a conference organised by the Historic Farm Buildings Group in association with the Rural Development Commission, RICS and English Heritage entitled *Old Farm Buildings in a New Countryside*, held at the RICS Westminster Centre, London, 29 November.
Saunders, M. (1991) Nonconformist chapels: the conservation challenge, *Ancient Monuments Society*, Vol. 35, pp. 163–85.
SAVE (1987) *Churches: A Question of Conversion*, SAVE Britain's Heritage.
SAVE (1988) *A Future for Farm Buildings*, SAVE Britain's Heritage.
SAVE (1990a) *Bright Future: The Re-Use of Industrial Buildings*, SAVE Britain's Heritage.
SAVE (1990b) *Historic Hospitals at Risk*, SAVE Britain's Heritage.
Smith, J. (1990) Paper concerning the problems of conversion presented by the founder of the Landmark Trust, Sir John Smith, at a conference entitled *The Conversion of Listed Buildings*, organised by the Historic Houses Association and held at the Royal Institution, London, 21 March.
Sutch, R. and Davey, K. (1993) Information Sheet No. 1, *Approach to Property Insurance and Insurance Valuations for Historic Buildings*, published as a supplement to the RICS Building Conservation Newsletter, No. 8, Summer.
Swallow, P., Watt, D. and Ashton, R. (1993) *Measurement and Recording of Historic Buildings*, Donhead.
Wainwight, M. (1994) Restoration of a Victorian church is becoming a virtual reality, *The Guardian*, 29 November, p. 5.
Weaver, M. (1993) *Conserving Buildings: A Guide to Techniques and Materials*, Donhead.
Weir, H. (1989) *How to Rescue a Ruin by Setting up a Local Buildings Preservation Trust*, Architectural Heritage Fund, pp. 25–49.

Chapter 7

CONSERVATION AREAS

Introduction

The first consideration of the idea of giving protection to areas can be traced back to the case *Iveagh v. Minister of Housing and Local Government* [1964] JPL 395 (see Chapter 2) in which the court held that a building might be of special architectural or historic interest by way of its setting as one of a group, although in this instance it was determined that the buildings were not of sufficient merit to warrant protection individually. However, the Court of Appeal concluded that a more general power was needed to protect matters of 'group value'.

In 1965 the Council of British Archaeology (CBA) published a *List of Historic Towns.* From an initial list of 660, 324 towns were selected which possessed buildings of architectural or historic importance and historic street-plans. The CBA's *Memorandum on Historic Towns* stated that the inclusion of any town within this list provided

... an argument for preparing for it a comprehensive survey of the historic environment, illustrating its layout, its historic buildings, its urban quality and any other special characteristics. This 'heritage plan' should ... form an obligatory part of the development plan process and should make specific provision for the conservation of features emphasized by the survey.

Moreover, 51 towns were singled out which the CBA stated were so important that proposals affecting their historic centres should be decided directly by the Minister of Housing and Local Government rather than by their LPAs (Burke, 1976).

However, due to relative inaction by the government, a Private Member's Bill, which was sponsored by Duncan Sandys M.P. (formerly Minister of Housing and Local Government and President of the Civic Trust) and passed with government backing, introduced the concept of *conservation areas.* During the passage of the Civic Amenities Bill the government gave some support to the CBA's recommendations. This was limited to the commissioning of four consultant pilot studies known as the *Four Town Reports* in 1966 (covering Bath, Chester, Chichester and York). These studies were designed to establish ways of halting the destruction of the fabric of historic centres. Before the publication of the individual *Four Town Reports* in 1968 the four areas were chosen for the first conservation area designations.

Section 1 of the Civic Amenities Act 1967 required LPAs to 'determine which parts of their areas ... are of special architectural or historic interest, the character and appearance of which it is desirable to preserve or enhance' and to designate such areas accordingly. Once designated, LPAs were required to pay 'special attention' to applications for new development within them and 'special scrutiny' regarding new buildings, alterations to existing buildings (particularly listed buildings), changes of use, street furniture and street patterns. Yet apart from the power to designate conservation areas no new powers of control were introduced.

A report entitled *Preservation and Change*, which was published by the Ministry of Housing and Local Government (MHLG) in 1967, expressed considerable concern at the rate of loss of buildings of quality, and with an eye for future policy initiatives discussed the need for a conservation policy, suggesting that the preservation of areas of architectural or historic interest should be compatible with change, particularly by directing new life into old areas (Reynolds, 1975). In 1968 LBC was introduced to strengthen protection measures on listed buildings but nothing was added in relation to conservation areas. However, the government's *Preservation Policy Group*, set up to coordinate and consider the *Four Town Reports*, recognised the growing importance attached by public opinion to preservation and conservation when it reported in 1970 (Ross, 1991). Yet the opportunity created by the consolidation and reform of planning legislation in 1971 added nothing further. Further encouragement was given to the development of conservation policy by the MHLG's report *New Life for Old Buildings* in 1971 and campaigning action by the Civic Trust at the same time (see Chapter 4). The first real change was subsequently brought about via the Town and Country (Amendment) Act 1972.

The 1972 Act introduced two matters of significance in relation to conservation areas. First, s. 10 permitted funds to be made available by way of grant or loan for building repair work connected with the promotion, preservation or enhancement of outstanding conservation areas. This power remains today under s. 77 of the P(LBCA) Act 1990 although the distinction of 'outstanding' was removed by Schedule 15, para. 27 of the Local Government, Planning and Land Act 1980. Secondly, a power was given to enable LPAs to prevent the demolition of unlisted buildings or groups of special architectural importance considered to be essential to the character of a conservation area. This power was limited by the need for a 'direction' for the protection of specific buildings. However, the Town and Country (Amenities) Act 1974, introduced via a Private Member's Bill by Michael Shersby M.P., strengthened this provision by providing an automatic control over the demolition of any unlisted building (excepting minor structures) without consent.

The 1974 Act, which was designed to make further provisions for the control of development in the interests of amenity, had further implications. The Secretary of State for the Environment was empowered to designate conservation areas himself (a power which, so far, has not been used). More significantly, LPAs were required to draw up and publish proposals for the preservation and enhancement of conservation areas, to submit such proposals to a public meeting and to have regard to any views expressed. Furthermore, additional amenity provisions were made including a temporary protection to all trees located within a conservation area not subject to a Tree Preservation Order and a power for the Secretary of State to make special provisions for the control of advertisements.

No other major changes were made in relation to the protection of conservation areas for a considerable amount of time. In fact the General Development Order made no specific mention in relation to conservation areas until the new order of 1988. However, the number of conservation areas designated since the Civic Amenties Act 1967 increased dramatically. In 1992 it was estimated that 4 per cent of the nation's building stock in about 7500 conservation areas were afforded a measure of protection by virtue of their location (Davies, 1992). By 1994 about 8000 conservation areas had been designated.

However, in the 1980s a number of contentious issues started to emerge in relation to conservation area policy. First, in relation to new development the meaning of the words 'preservation', 'enhancement', 'character' and 'appearance', as found in the designation power for conservation areas and associated provisions, came under considerable scrutiny following the case of *Steinberg v. Secretary of State for the Environment* [1989] 2 PLR 9 (Millichap, 1990; Stubbs and Lavers, 1991). Professor Steinberg succeeded in overturning the Secretary of State's decision to grant planning permission for development in a conservation area by having it quashed on the ground that there was a difference between whether the proposed development would 'harm' the character and appearance of the conservation area (the criterion applied by the Secretary of State's inspector) and the appropriate conservation area criterion (i.e. the special attention to be paid to the desirability of preserving or enhancing the character and appearance of the area).

The second area of contention concerned the considerable growth in the number of conservation area designations and whether this had watered down the original concept of the policy. In particular, concern was expressed by the *English Historic Towns Forum* and other conservation groups regarding the fact that the features which make up the character and appearance of designated areas have frequently not been recorded, that the cumulative effect of many minor alterations to unlisted buildings beyond planning control have led to a reduction in the quality of areas, and that the primary reason for designation has often been perceived as securing conservation finance (EHTF, 1992).

However, these issues have come under considerable scrutiny. The development of a new approach to conservation area policy was carried out over a period of about three years with the eventual publication of PPG 15 in September 1994 after considerable consultation. This heralds a new era for conservation area policy which includes the CAP initiative, new controls and greater clarity on policy issues.

Designation

The duty placed on LPAs to 'determine which parts of their area are areas of special architectural or historic interest' and to designate them as conservation areas is now contained in s. 69 of the P(LBCA) Act 1990. This duty has to be reviewed 'from time to time' in relation to existing conservation areas and further parts of their area.

The power to designate conservation areas is principally carried out by district or metropolitan planning authorities or London Borough Councils as LPAs within London. However, county planning authorities have a power to designate, though this is not a duty (Moore, 1994b), outside a National Park but the county LPA must first consult the relevant district LPA. In London, English Heritage may also make designations but must first consult the relevant London Borough Council. The Secretary of State for National Heritage also has a power to designate subject to consultation with the relevant LPA. However, para. 4.8 of PPG 15 indicates that this power will only be exercised in 'exceptional' cases such as where the area is of more than local interest, or where a LPA is reluctant to designate an area due to its ownership of important buildings in the area despite the fact that there is a clear threat to the character or appearance of the area.

Conservation areas vary in size, scale and character and are usually designated for townscape quality in its broadest sense, not just relating to individual buildings in an area. Historic town and city centres are usually designated as one or a number of conservation areas. Squares, terraces and smaller groups of buildings, model villages and suburban areas have been designated. About half the total number of designations are located in villages and rural settlements. All LPAs have designated at least one conservation area but some have chosen to designate many areas individually covering a small area (Pearce *et al.*, 1990) and one authority (Chichester) has made over 80 designations (Moreton, 1991). In London substantial parts of some boroughs now lie within conservation areas, for instance over 76 per cent of the City of Westminster has been designated (Andreae, 1992). Examples of the wide variety of designations include Hampstead Garden Suburb, inter-war housing estates in the London Boroughs of Havering and Harrow, the historic city centre of York, a former industrial area of Birmingham known as the Jewellery Quarter, the whole of the Settle to Carlisle railway stretching over the counties of North Yorkshire and Cumbria, a Victorian cemetery in Bradford, and the workers' village of Silver End in Essex.

Designation procedures

The statutory procedures for desigation are relatively simple. Moreover, LPAs have traditionally designated conservation areas through the resolution of council members by reference to a map showing the boundary of the area. A map has often been published as part of the statutory publicity requirements. Under s. 70(8) of the P(LBCA) Act 1990 a designation (or a variation in the boundary or the cancellation) of a conservation area must be given notice in the *London Gazette* and at least one local newspaper. Owners and occupiers of property lying within a conservation area do not have to be specifically notified and there is no right to object to a designation. At the same time, some LPAs have chosen to notify individuals and in any event public involvement in conservation area policies is given official encouragement as designation imposes a duty upon the LPA under s. 71 of the P(LBCA) Act 1990 to formulate and publish proposals for the preservation and enhancement of any conservation area within their district, and to submit them to public consultation.

A designation is also registrable as a local land charge (s. 69(4)) and the Secretary of State for National Heritage (and English Heritage) must be notified (s. 70(5)), but his consent is not necessary.

Problems associated with designation

The process of review of conservation area policy has undoubtedly led to a considerable growth in the number of conservation areas since the relevant provisions were first enacted in 1967. However, since this time there has been increasing concern that the intentions of the legislation have become devalued.

The main criterion which has been used over the history of conservation area policy in deciding whether to designate an area as a conservation area has been to determine whether an area has 'special architectural or historic interest, the character or appearance of which it is desirable to preserve or enhance'. For the majority of time there has been no precise specifications to aid LPAs in deciding whether to designate apart from brief indicators, most recently found in para. 54 of DoE Circular 8/87, that they may be 'large or small, from whole town centres to squares, terraces and smaller groups of buildings', or that they could be 'centred on listed buildings' or that 'pleasant groups of other buildings, open spaces, trees, a historic street pattern, a village green or features of historic or archaeological interest may contribute to the special character of an area'. Furthermore, some commentators have questioned the extent to which comprehensive enhancement plans have developed beyond the stage of formulation and publication, or the extent to which conservation area advisory committees, which the government first asked LPAs to set up in 1968 and has subsequently given numerous reminders in circular advice since, have been been set up to assist in policy formulation (Bloxsidge, 1975; Moreton, 1991; Millichap and Judd, 1991; Ross, 1991).

Although the designation of many conservation areas will most likely have been subject to some form of initial character appraisal in order to determine matters of architectural or historic interest worthy of preservation and enhancement, with many designated areas being 20 or more years old it would appear that the guiding influence of such appraisals has, in some instances, become subsumed into more general local plan policies. Moreover, the updating of conservation area policy through boundary changes to existing designations or the creation of new areas has sometimes been more of a response to particular development control problems, such as a means to control demolition, rather than arising from a specific detailed appraisal of architectural and historic character (Moreton, 1991). In turn this has not provided a clear basis for development control decisions as without a clear statement on matters of character it has become more difficult to determine how, for instance, new development in an area would affect the 'special architectural or historic character' of an area.

These problems have been compounded by the fact that only the demolition of unlisted buildings, and not alterations to them, have been subject to control. Evidence of the gradual decline in the appearance of many conservation areas through permitted alterations has been provided from a number of sources and were highlighted in the 1992 report *Townscape in Trouble: Conservation Areas – The Case for Change* (EHTF, 1992) and the *London Conservation Areas Conference* organised by the Georgian Group in 1992.

LPAs have been able to apply to the Secretary of State for the Environment for an 'Article 4 direction' under the Town and Country Planning General Development Orders (GDO) (see Appendix 6) to bring certain items of 'permitted development' under planning control. In relation to conservation areas many damaging alterations to single family dwellinghouses are permitted development and do not require consent unless development rights are withdrwan by an 'Article 4 direction'. These include the following items under Schedule 2 of the 1988 GDO:

- *Part 1, class A(1)* Extension to a single family dwelling by up to 50 m^3 or 10 per cent (whichever is the greater) of the original size subject to certain caveats regarding overall height, proximity to boundary, position of extension in relation to a highway and percentage of overall site. (Under part 1 class A(2) permitted development rights are not given in relation to class A(1) for land included in Article 1(5) land, which includes conservation areas, where it would consist of cladding of any part of the exterior with stone, artifical stone, timber, plastic or tiles.)
- *Part 1, class B* Enlargement of a dwellinghouse by an addition or alteration to a roof subject to certain size tolerances and caveats regarding the existing height of the building and plane of the slope.
- *Part 1, class C* A roof can also be altered by replacement of roof covering and insertion of roof lights provided they are flush with the roof material.

- *Part 1, class D* Erection of a porch outside any external door of a dwelling-house.
- *Part 1, class F* Provision of a hard surface within the curtilage of a dwelling for any purpose incidental to the enjoyment of the dwelling.
- *Part 1, class H* Provision of a satellite antenna subject to certain size and height restrictions.
- *Part 2, class A* (any building including a dwellinghouse) erection, construction, maintenance, improvement or alteration of a gate, fence, wall or other means of enclosure.
- *Part 2, class C* (any building including a dwellinghouse) painting of the exterior of a building.

However, the use of 'Article 4 directions' to control alterations to dwelling-houses in conservation areas has not proved to be a satisfactory method of preserving the character of unlisted buildings in conservation areas for two reasons.

First, the Secretary of State has been reluctant to approve some 'Article 4 directions'. Designation as a conservation area is insufficient in itself. It is necessary to demonstrate that an area has a certain quality and importance and that there is a special need for a direction (English Heritage, 1993). This situation is different to the position regarding permitted development rights associated with listed buildings as LPAs are allowed to remove these rights without recourse to the Secretary of State.

Secondly, some features of buildings which may contribute to the character and appearance of an area have been allowed to be removed because there has been uncertainty under planning legislation as to whether the alteration constitutes 'development' (as defined under s. 55 of the principal planning Act) or would otherwise 'materially affect the exterior of a building', which would normally require that a planning application be submitted. In particular, Schedule 2, part 1, class A of the 1988 GDO (concerning development within the curtilage of a dwelling house), which permits enlargement, improvement and other alterations, has not specifically mentioned such matters as windows and doors. The proportions of glazing and joinery, and profiles and mouldings of external doors and windows have been recognised as being fundamental to the appearance of any building (Fig. 7.1) (MacQueen, 1992). However, while English Heritage has attempted to combat the loss of character from the blocking out of existing doors and windows, the creation of new openings and the drastic effects of uPVC replacements through its *Framing Options* campaign, the majority of problems have remained. Similar problems have arisen in relation to the removal of architectural details such as cornices, porches and chimneys.

Thus there has been a dilemma as 'Article 4 directions' for conservation areas have not always been approved because the character has been eroded by the loss of details which have not been regulated. This led to a review of conservation area policy during the early 1990s. Yet while an intention to

Fig. 7.1 Replacement windows in unlisted houses within the Cromford Conservation Area, Derbyshire.

introduce new controls over alterations to unlisted buildings was announced by the Department of the Environment in 1994 and 1995 (see below), two cases have questioned the power of LPAs to designate conservation areas.

In the first case of *R. v. Canterbury City Council, ex parte Halford* [1992] 2 PLR 137, an extension to an existing conservation area at Barham, Kent, had been questioned by Kent County Council on the grounds that it would diminish the value of the original area which had a distinct character. While the city council subsequently decided to approve the extension (to include the setting of a village as well as the already designated village), the owner of a field incorporated within the new area boundary challenged this decision following the refusal of permission to develop. The basis of challenge was that irrelevant matters had been taken into account in that the area in question was to include an 'important tree-lined feature', 'fine views of the church and village centre', a 'meadowland', that no buildings or structures existed on the field in question and that the trees on it were already protected by tree preservation orders (TPOs).

The extension was quashed because the report that had recommended the designation to committee members had not drawn attention to existing TPOs and it was found that this could have influenced the decision to extend. Nevertheless, while no precedent had established the proper approach to be adopted in designating conservation areas, McCullough J. determined that the intention of s. 69(1) of the P(LBCA) Act 1990 was to indicate to LPAs that in determining an 'area' as being of 'special architectural or historic interest' the approach should be to consider the whole of an area of land as an entity, meaning that not every part of the area need have something on it which is of special interest. Of note he referred to the fact that Circular 8/87, para. 54 – the relevant planning guidance in force at the time – had emphasised that conservation areas may be widely drawn.

In the second case of *R. v. Swansea C.C., ex parte Elitestone Ltd* [1992] JPL 1143, a similar conclusion was drawn as McPherson J. considered that the LPA had a wide discretion to reach its own conclusions as to which parts of the area had special interest. In this case a settlement of self-built chalets had been included in a conservation area on the basis that they were considered to be an example of plot land developments, unique to the nation's built heritage. Thus an application to redevelop the site was rightly refused as the chalets, their site and tenure system were found to be of historic interest and worthy of preservation and enhancement.

Both of these cases have pointed to consideration of an area as a whole, whether or not the area includes buildings or structures. Moreover, the wide discretion exercisable by LPAs in this context has meant that it is difficult to challenge a decision to designate on its merits unless the LPA has taken into account irrelevant considerations or failed to take account of something which is relevant (Mynors, 1993a). Yet some commentators have argued that the designation process has been open to abuse – that LPAs have used conservation area designations as a means to apply stricter planning controls, particularly as it has been possible to designate areas outside the normal development plan process, and as some areas have been designated without recourse to a fully detailed assessment of the character or appeareance of the area (Graves and Ross, 1991). Or, in another sense, even where character appraisals exist they have frequently not been used for development control purposes (Moreton, 1994).

However, the position has changed since the priority of the development plan for forming planning decisions was introduced through the amendment of the principal planning Act by s. 54A and the issuance of new policy guidance through PPG 15.

New designation guidance

In 1993 English Heritage issued a guidance note on *Conservation Area Practice* which was published in the context of the emerging consultation draft of PPG 15 (English Heritage, 1993). It highlighted the fact that s. 54A has emphasised the importance of including firm conservation area policies in local plans which should be based on a clear definition of what constitutes 'special architectural or historic interest' in every case.

Thus, in relation to proposed new designations, LPAs have been advised to judge 'special architectural and historic interest', with the assistance of specialist professional advice, against locally related criteria. Moreover, the need for real quality or 'specialness' was emphasised in terms of the local or regional context. Furthermore, it was suggested that designation would be best achieved in connection with the preparation or review of a district-wide local plan. In the case of existing designations the guidance suggests that there should be district-wide reassessments to see whether areas still have 'special interest' and a decision made to review boundaries or even de-designate those areas which are no longer considered to be 'special'. At the same time this process of review would also allow for the redrawing of boundaries where they have been too narrowly drawn or which have omitted areas, such as Victorian and Edwardian developments, which may now be considered to be important.

More significantly, because of the importance of establishing the special interest in local plan policies in order to provide a clear basis for development control decisions (as well as for formulating proposals for the preservation and enhancement of an area) the guidance has attempted to lay down a checklist of issues which may be used for an area's appraisal. These include:

- the origins and development of the topographic framework of the area
- the archaeological significance and potential of the area, including any scheduled ancient monuments
- the architectural and historic quality, character and coherence of the

buildings, both listed and unlisted, and the contribution which they make to the special interest of the area
- the character and hierarchy of spaces, and townscape quality
- the prevalent and traditional building materials
- the contribution made by greens or green spaces, trees, hedges and other natural or cultivated features
- the prevailing or former uses within the area and their historic patronage, and the influence of these on plan form and building types
- the relationship between the built environment and landscape or open countryside, including the significance of particular landmarks, vistas and panoramas where appropriate

The guidance also makes particular reference to the need for assessment of the contribution which unlisted buildings may make to an area. This may be in terms of such matters as age; style; materials; phase of development; historical association with particular features such as a park or burgage plot, or with people or events; traditional function or former use; group value; and any landmark quality.

Also important in the assessment is the identification of any neutral areas which neither diminish nor enhance the area and, more significantly, any negative factors which detract from the character of the area. Furthermore, the guidance emphasises the need for an on-going dated photographic record of an area to monitor change, the physical condition of buildings and the progress of enhancement schemes. The progress of Conservation Area Partnership regeneration schemes, introduced in April 1994, should now be added to this list.

When PPG 15 was issued in September 1994 specific mention was made of English Heritage's guidance note. Moreover, the new policy statement made a significant shift from the position of its predecessor, Circular 8/87, in emphasising the importance of a LPA's justification for designation by the assessment of an area's special interest, character and appearance as these are now factors which the Secretary of State will take into account when considering appeals against refusals of conservation area consent (CAC) for demolition and refusals of planning permission (para. 4.5). It has also been recommended that the process of assessment and the formulation of proposals for areas should involve extensive public consultation. Separate from this, development plans have been identified as the appropriate mechanism for designating or reviewing existing conservation areas and to provide a policy framework for weighing decisions on applications for planning permission and CAC (para. 2.9). Furthermore, para. 4.4 highlights the fact that the more clearly the special interest of an area is identified in the process of designation (or review), the sounder will be the basis for local plan policies and development control decisions.

Thus an attempt has been made to resolve the weaknesses surrounding the designation of conservation areas. The new policy is part of a wider examination of planning policy to ensure 'sustainable development' by the safeguarding of key environmental resources, including the built heritage, which has been the subject of further examination jointly by English Heritage, English Nature and the Countryside Commission through the publication of further guidance entitled *Conservation Issues in Stategic Plans* in 1993. However, there will be an inevitable delay to the implementation of new guidance on conservation area designation in the development plan system as it moves towards a nation-wide coverage of local plans on a district-wide basis, particularly as the administration of local government is commencing a period of reorganisation.

Nevertheless, PPG 15 marks a new era for conservation area policy which should eventually lead to improvements in development control practice and a clearer basis for preserving and enhancing the character and appearance of conservation areas.

Despite the new emphasis of the development plan and the fact that it is for LPAs to decide on the level of detail of such plans, the Secretary of State has been known to direct against excessive detail and inflexibility in the approval of conservation area policies in plans. The object should be to preserve and enhance through *managed change* and not to establish inflexible rules which are too 'prescriptive, onerous and restrictive'. This was the case in *R. v. Secretary of State for the Environment, ex parte Mayor and Burgess of the London Borough of Islington* [1995] 121.

Preservation and enhancement of conservation areas

A duty is placed on LPAs by s. 71 of the P(LBCA) Act 1990 to formulate and publish proposals for the preservation and enhancement of conservation areas. These should be submitted to public scrutiny via a public meeting and consideration by a wide range of community interests, including residents, business interests, amenity societies, highways authorities and public utilities. Some LPAs have also set up *Conservation Area Advisory Committees* (which the government has recommended since 1968), made of a cross-section of community interests and nominations from national bodies such as the national amenity societies and the Civic Trust, to assist in this process.

Thus not only is character appraisal important in the first instance, action to protect the special interest is necessary following designation. Here again the local plan is considered the most appropriate mechanism to integrate conservation policies with wider policies for an area. Of particular importance is the treatment of roads, pavements and public spaces; the use of grant aid to maintain townscape buildings; and the use of planning powers to control development and manage areas, including the development of policies to ensure the use of vacant premises which contribute to the townscape quality. In some cases this may be achieved through the setting up of a

Conservation Area Partnership to assist in the economic regeneration and enhancement of an area.

Townscape improvements

Apart from preserving and enhancing the character and appearance of a conservation area by sympathetic new development and planning control, physical improvements to the streetscape are also a means to assist this aim. This involves consideration of the floorscape, street furniture and signage, architectural features and traffic management and the use of property.

Vacant premises may detract from the qualities of a conservation area as the lack of income prevents repair and maintenance. Consideration has been given to the problem of unoccupancy particularly in designated areas of town or city centres. The *Living Over the Shop* scheme brought encouragement for the re-use of upper floors until central funding was cut. Nevertheless, LPAs have been encouraged to continue such schemes through para. 4.11 of PPG 15 and through the UK Strategy on *Sustainable Development* which, apart from encouraging environmentally led planning in a general sense, seeks to ensure that the built heritage is sustained. Inevitably this requires imaginative solutions and one of the best opportunities for bringing empty upper floors of premises back into use has been recognised through allowing conversion to residential use with the added benefit of keeping areas alive at all times of the day. Housing Associations have considerable scope to secure such conversions. Moreover, the housing renovation grant system may assist this process if LPAs develop policies to secure better use of vacant premises.

Funding is an important aspect of any enhancement scheme. Section 77 of the P(LBCA) Act 1990 has been the main source of grant aid from English Heritage for conservation areas (although LPAs make grants or loans available for repair works to buildings of achitectural or historic interest, whether listed or not, through s. 57 of the Act). Expenditure permitted under s. 77 must make a significant contribution towards preservation and enhancement and has usually only been provided where a particular programme of work has been invited; where there is a Town Scheme or a Conservation Area Partnership arrangement; where the LPA has initiated a conservation scheme in a particular place such as a terrace or a square; where works to a building or complex of buildings are to act as a catalyst for further action; or where buildings have been subject to statutory repair notices.

There are two eligible items for grant aid to buildings in conservation areas under s. 77. First, those which contribute to *keeping up the structure* of a building if this directly relates to maintaining the character of an area. These include general structural repairs, specialist treatments and reinstatement works for items removed in the course of repair work. Secondly, external items, visible to the public, which contribute to the *physical preservation or enhancement of the conservation area* may be eligible. These include roof repairs, chimney repairs, leadwork, rainwater goods, brickwork and stonework repairs, repointing, joinery work, external cleaning and replacement or reinstatement of special architectural features. An emphasis is put on the use of natural original materials and repair rather than replacement, though the provision of accurate replicas of original designs and materials in the case of such matters as roofing materials, rainwater goods, doors, windows and other external joinery which have been altered by modern replacements may be allowed if they contribute to the character of the building and/or conservation area. For example, many LPAs have published design guides to encourage, sometimes with the support of financial aid, the reintroduction of traditional shop front designs which can make an important contribution to the character and appearance of an area.

In addition, environmental works such as paving in appropriate natural materials and appropriate street furniture, and external works associated with grant-aided projects may be eligible. The latter may include works to retaining walls or boundary features which enhance the setting of a building. Environmental grants are given on the premise that the spaces between buildings should not be separated from the buildings themselves as they may form important elements of the overall character of a conservation area. However, the finite sources of funding means that conservation grants for this purpose are second in priority to actual repair works to buildings.

In many cases county authorities with responsibility for highways have been prepared to pay for environmental streetworks in association with district authorities (Lillford and Dadson, 1994; Grover and Lewis, 1994). Forward-thinking county authorities have developed guidance notes on conservation areas, particularly where highways and planning matters have been administered from the same department. In some instances English Heritage has collaborated with some authorities to demonstrate the advantages of a correct approach using local materials in a low key manner which, nevertheless, reinforces local identity (Figs 7.2 and 7.3) (Davies, 1991). This has arisen because some historic areas have been damaged by misguided 'improvements' which have taken the form of a 'wall-to-wall carpet' of durable, artificial materials and in inappropriate colours, such as the introduction of modern red tile finishes or other standardised block finishes in pedestrianisation schemes (English Heritage, 1993; Booth, 1994). Moreover, in response to this problem English Heritage produced a guidance leaflet in 1993 entitled *Street Improvements in Historic Areas.* Others have argued that the importance of researching historic buildings before carrying out repairs should be matched in the context of enhancement of historic street finishes, particularly as there remains a wealth of building stone suitable for paving purposes (Green, 1994).

While many items of historic street furniture – such as red telephone boxes, post boxes, bus shelters, and memorial features – have now been lost, an opportunity may be made to retain what is left through listing or

Figs 7.2 & 7.3 Street improvements in the historic town of Barnard Castle, County Durham. The scheme uses natural stone paving, stone setts, and appropriate street furniture; the sensitive treatment has been extended to manhole covers.

otherwise insuring that they are maintained and repaired. Careful use of new street furniture is also important. There are now many firms manufacturing street furniture items in traditional designs. However, there is also a danger of inappropriate use of these features, such as over-elaborate Georgian style lanterns in a village setting (Harris, 1994). In other cases the use of modern street lighting fittings may be controversial and can detract from the character and appearance of a conservation area.

Transport and traffic management is also a key factor in the preservation and enhancement of conservation areas. Moreover, s. 5 of PPG 15 is devoted to this issue. In addition to the safeguarding provisions of the Transport and Works Act 1992 (see Chapter 3) the planning guidance emphasises the need for local highway and planning authorities to be cognizant of the wider costs of transport choices, including their potential impact on the historic environment. Moreover, PPG 13 (Transport) issued in 1994 was one of the first revisions of planning policy which emphasised the need for sustainable development by encouraging means of travel which have less environmental impact.

Some small historic towns have been damaged through heavy lorries and cars causing vibration and pollution apart from the visual intrusion in the historic setting. While many new routes have been built to bypass and safeguard these centres, other opportunities to reduce the penetration of vehicles in historic centres are being created by *park-and-ride* schemes such as have been introduced in York, Cambridge and Oxford. Pedestrianisation, traffic-calming and other vehicle restrictions are also now being encouraged through changes in government policy and the advice of specialist bodies. For example, the English Historic Towns Forum has produced two relevant guidance documents entitled *Park and Ride Good Practice Guide* and *Traffic Measures in Historic Towns*, while the Department of Transport has issued advice on a range of traffic-calming measures which may be introduced. Moreover, the Highways (Traffic Calming) Regulations 1993 have enabled local authorities to be flexible in the use of a variety of traffic calming measures including road humps, chicanes, build-outs and islands. Although these can be difficult to integrate into older townscapes, local authorities are increasingly using consultants to tackle the problem of vehicle penetration and reintroducing natural materials into the streetscape with considerable success (Lawton, 1994). Further support for this approach has been suggested following the New Roads and Street Works Acts 1991 and the publication of PPG 8 (Telecommunications) in 1992 with particular reference to the need for highway authorities and public utility companies to comply with policies on pavement surfaces set out in local plans, or for them to agree a code of practice with LPAs.

Often streetscape improvements can act as a significant catalyst in restoring the quality of historic areas, inducing repair and re-use of properties as the environment improves with consequent economic benefits. However, planning control also plays an important part in safeguarding buildings of architectural or historic interest and through the control of development to ensure that new buildings preserve and enhance the character and appearance of conservation areas.

Planning control in conservation areas

Under s. 72 of the P(LBCA) Act 1990, LPAs must give special attention in the exercise of planning control to the desirability of preserving and enhancing the character or appearance of conservation areas. While control may be regarded as a negative function of the planning system the overall aim of this section is positive. The status now given to development plans via s. 54A of the principal planning Act has made it important for LPAs to set out their strategy for planning control in conservation areas in local plans. As the system of district-wide local plans gradually comes into being, this should ensure that all conservation areas are subject to detailed assessment and proposals which aim to ensure that the desire of s. 72 is achieved.

Paragraph 4.15 of PPG 15 emphasises the need for controlled and positive management of change. Proposals for the demolition of buildings must be viewed against the contribution they make to the architectural and historic interest of an area, including the wider effects of loss to the surrounding area and the conservation area as a whole. Likewise, new provisions on alterations to unlisted buildings will extend the control over minor changes which cumulatively can have a detrimental impact on a designated area. Gap sites or buildings which make no contribution to the area may present opportunities for sympathetic development to keep an area alive in social, economic and aesthetic terms. Other aspects of environmental control may also be exercised in relation to trees and advertisements to ensure the aims of s. 72.

Demolition control in conservation areas

Control over the demolition of most buildings in conservation areas is provided by s. 74 of the P(LBCA) Act 1990. The need for CAC does not apply to listed buildings, scheduled ancient monuments, or any ecclesiastical building which is in ecclesiastical use and any other matter which the Secretary of State directs. Presently 11 classes of building have been specified in a Secretary of State direction, and these are one of the remaining issues which for which DoE Circular 8/87 remains in force (see para. 97). While PPG 15 has not clarified directions for the handling of applications for LBC and CAC, it is intended that these will be updated (see Appendix 7).

In determining applications for CAC, LPAs are required to have special regard to the statutory duty contained in s. 72 (see above). This is the prime consideration and, as with applications for LBC, the decision-making process should not be affected by the requirement regarding the primacy of the development plan as applied to applications for planning permission under s. 54A of the principal planning Act as the P(LBCA) Act 1990 makes no mention of this. While legal commentary has suggested that, under the normal principles of administrative law, the policies in a development plan may have to be regarded as a relevant consideration in determining matters in relation to listed buildings and conservation areas, the view has been expressed that the most likely interpretation is that Parliament only intended s. 54A to apply where legislation expressly required it (Purdue, 1994). Moreover, in *St Albans D.C. v. Secretary of State for the Environment and Allied Breweries Ltd* [1993] JPL 374, Sir David Widdicombe Q.C., sitting as Deputy Judge, accepted that s. 54A did not apply to applications for CAC under s. 74 of the P(LBCA) Act 1990.

Paragraph 4.27 of PPG 15 states that there should be a general presumption in favour of keeping buildings which positively contribute to the character and appearance of a conservation area. Applications for CAC should be judged against the same criteria as applications for the demolition of listed buildings. Through this process there will be instances where some buildings are judged as making little contribution to an area and it may be possible to grant CAC. Neverthess, if demolition is to be permitted there is another important matter to consider – what will happen to the site following the removal of a building? It has been recognised that leaving an unsightly gap may be detrimental to an area. Moreover, there is also the consideration of the form of development that will replace a demolished building and whether this will enhance or detract from the character and appearance of an area.

To deal with this latter position it is often deemed appropriate to grant CAC subject to a condition that demolition will not take place until a contract for works of redevelopment has been made and planning permission also granted. In the case of a proposal to demolish a listed building within a conservation area this position is emphasised by s. 17(3) of the P(LBCA) Act 1990. Moreover, as indicated in Chapter 2, the case of *Davis v. Secretary of State for the Environment and Southwark L.B.C.* [1992] JPL 1162 has confirmed that the relevant statutory provisions are interdependent in that a 'suitable redevelopment scheme' must meet the requirements for granting LBC for demolition and planning permission for new development under both protection regimes. Thus, while the primacy of the development plan is not required to be taken into account where only demolition is being considered, the provisions of s. 54A will bite where a proposal for demolition is subject to conditions regarding redevelopment, or otherwise is part of a scheme of new development.

A potential problem area for interpretation regarding the demolition of buildings in conservation areas arises in the same way as with listed buildings over the definition of the word 'demolition' (see the *Shimizu* case in Chapter 2 and *Kent County Council v. Secretary of State for the Environment* [1994] EGCS 91).

New development in conservation areas

Where an application for planning permission for *any* land is made and, in the opinion of the LPA, would affect the character or appearance of a conservation area, the LPA is required by s. 73 of the P(LBCA) Act 1990 to give publicity to the proposal. The publicity requirements which apply are the same as apply to applications affecting the setting of listed buildings under ss. 67(2)–(7) of the P(LBCA) Act 1990, including the need for a period of 21 days to allow for representations to be submitted which, in turn, must be considered in the decision-making process.

Advice provided by English Heritage regarding the consideration of proposals for new buildings suggests that LPAs should look at the appropriateness of: the overall mass or volume of a proposed building; its scale and size with particular reference to windows, doors, floor heights, and units; and whether it will be harmonious or complementary to its context in terms of neighbouring buildings, materials and its visual impact on the townscape or street scene (Figs 7.4 and 7.5). Design or development briefs may be utilised for sites identified for new development or the replacement of features, including buildings, which detract from the character or appearance of an area (English Heritage, 1993). Furthermore, according to guidance contained in para. 4.18 of PPG 15, LPAs may more usually require detailed plans and drawings of any proposed new development, including elevations to illustrate the building in its setting, before considering a planning application.

The *Steinburg* case initially had a considerable effect on the way in which applications and appeals over development proposals in conservation areas were decided. In many cases it was widely interpreted as requiring refusal of planning permission unless it could be shown that the development proposed would preserve or enhance the character and appearance of an area. However, in *Bath Society v. Secretary of State for the Environment* [1991] JPL 663, Glidewell L.J. attempted to provide authoritative guidance regarding the approach that should adopted in relation to the duty contained in s. 72 of the P(LBCA) Act 1990 in the case of applications for new development. The following six propositions were put forward:

1. If permission be sought for development on a site not within a conservation area, the ordinary presumption in favour of permission as indicated in para. 15 of PPG 1 would apply, i.e. that it would not cause demonstrable harm to interests of acknowledged importance. (This is now superseded by revised guidance in the light of s. 54A of the principal planning Act.)
2. If the site is within a conservation area there are two statutory duties to perform – to have regard to the development plan and any other material considerations, and to pay special regard to the desirability of preservation and enhancement under s. 72.
3. In a conservation area the provisions of s. 72 should be the first consideration of the decision-maker and should be regarded as having considerable importance and weight.

Figs 7.4 & 7.5 Historic brick buildings in Marlborough, Wiltshire. The entrance to a new courtyard retail development which, although of modern design, has attempted to harmonise with the existing historic fabric in scale and materials.

4. If the decision-maker decides that the development would either preserve or enhance the conservation area this should be a major point in favour of giving permission for new development.
5. If a development proposal would both enhance and cause detriment to a conservation area, these two factors should be weighed against each other.
6. If development would neither preserve nor enhance then it would be almost inevitable that it would cause harm to the character and appearance of a conservation area. This should be a matter of considerable weight and importance. The decision-maker should only grant planning permission if there was some advantage or benefit which outweighed the test of s. 72 and the detriment which the development would cause.

One month after the decision in the *Bath Society* case the Court of Appeal, under a different constitution, further reviewed the position in the case of *South Lakeland District Council v. Secretary of State for the Environment and Carlisle Diocesan Parsonages Board* [1991] JPL 654. The court considered that the view expressed in proposition 6 was too rigid. In other words, if a development does not preserve a conservation area it would be wrong to assume that it would cause harm to it – it could have a neutral effect. This view was later confirmed by the House of Lords [1992] 1 PLR 143. Lord Bridge stated that planning policy should not be read too rigidly when it was difficult to see any rational justification for this. Thus it was concluded that the proposal to build a new vicarage in the grounds of the existing vicarage located in a village conservation area, which would largely be screened from view by a stone wall, shrubs and trees, would have little effect on the character and appearance of the area.

Since the *Bath Society* and *South Lakeland* cases, the provisions of s. 54A of the principal planning Act have come into force. The focus placed on the primacy of the development plan now raises the question of whether Glidewell L.J.'s third proposition that, of the two duties to perform (as highlighted in the second proposition), the consideration under s. 72 is still the first consideration for the decision-maker. This was considered in the *St Albans* case, the first scrutiny by the courts in relation to s. 54A which also happened to relate to a proposal for new development in a conservation area. Here Sir David Widdicombe Q.C. stated that it was not necessary for the decision-maker (in reference to the decision of the planning inspector to grant planning permission on appeal) to expressly refer to the development plan policies as long as they had been taken into consideration in the decision (Mynors, 1993b).

In this case the two conflicting statutory duties of s. 54A of the principal planning Act and s. 72 of the P(LBCA) Act 1990 were linked by the fact that the local plan set out the relevant policy to be applied in conservation areas. The proposal was to replace a nineteenth-century public house with a two-storey office building. The public house had been included in the conservation area 'broadsheet' as one of the buildings of local interest which contributed to the character of the area. Nevertheless, it had been extended in an unsympathetic manner and the inspector had concluded that it did not, in fact, make a significant contribution to the local scene. The office proposal involved a nominal departure from the development plan policy which, even allowing for the s. 54A presumption in favour of the plan, would have not defeated the proposal as there were material considerations for allowing the development.

More significantly, the Deputy Judge stated that the inspector had been 'rightly concerned with the requirements of s. 72' while also stating 'that does not mean that the development plan can be ignored'. The inference here being that s. 72 takes precedence over s. 54A. This position is mirrored by the *Heatherington* case (see Chapters 2 and 3) in which, although it was mainly concerned with alternative use schemes in relation to a listed building (also located in a conservation area), there was a duty to balance the statutory duties under s. 66 of the P(LBCA) Act 1990 and s. 54A. Moreover, the case specifically referred to the *Bath Society* and *South Lakeland* cases and drew the conclusion that, despite the slight difference in the wording of the duties expressed in ss. 66 and 72, there was no significance attached to this difference. In a more general manner, while s. 54A has greatly increased the importance of the development plan in relation to development decisions, there will be situations in which it will be possible to depart from it. Thus, while it will usually be regarded that the development plan has a higher status, there is still flexibility in the planning system if there are convincing reasons for departing from it (Purdue, 1994).

This was found to be the case in *Chorley and James v. Secretary of State for the Environment* [1993] JPL 927. Here the LPA had granted CAC for the removal of an unattractive asbestos shed located in a rural conservation, but had refused planning permission for its replacement by a new dwelling. This decision was confirmed at appeal on the basis that the proposal breached the development plan policy relating to dwellings in the countryside. However, the Deputy Judge stated that the inspector had failed to consider the character of the conservation area as a whole and, in particular, whether the harm in terms of the breach of residential policy outweighed the gain in terms of appearance – what the appellant argued to be 'the overall package' being offered.

Thus, while the *St Albans* case emphasised the importance of the development plan in coming to a decision, the material consideration that may be presented by consideration of s. 72 means that the decision-maker must balance both statutory duties. Moreover, the *Chorley* case indicates that development plan policies should not be slavishly applied where there are strong reasons for coming to a contrary decision. As the development plan system is gradually moving towards a system of district-wide local plans, this emphasises the need for the formulation of suitable policies concerning the preservation and enhancement of designated conservation areas which have been devised through detailed character appraisals. This is also important in the context of sites which neighbour conservation areas as the development of such sites may affect the setting, or views in or out of an area (Martin, 1995).

Control over alterations

In order to resolve the problem associated with 'Article 4 directions' and the fact that some alterations to buildings located within conservation areas have been outside planning control, the Secretary of State for National Heritage announced that the government intended to introduce new controls over external alterations to domestic buildings in conservation areas, including such matters as new doors, windows, roofs and frontages (DNH, 1994).

However, such a change requires Parliamentary approval and the delay in the formulation of an amendment to the GDO meant that the proposed changes to planning control could not be indicated in September 1994 when PPG 15 was issued. Nevertheless, para. 4.22 of PPG 15 further specifies that LPAs will be able to make directions to withdraw permitted development rights for certain prescribed items which merit additional protection in this context without the consent of the Secretary of State once the relevant legal authority is given, and provided that the proposals are published and consideration is given to representations made by local people. Furthermore, para. 4.23 specifies that the withdrawal of permitted development rights outside the new categories will continue to require an 'Article 4 direction' for which the Secretary of State's approval will continue to apply.

In November 1994 a consultation paper entiled *'Permitted Development' Rights for Dwellinghouses in Conservation Areas* was issued jointly by the DNH and the DoE. This indicated that amendments would be made to 'Article 5' of the GDO which gives LPAs the power to make 'directions' to withdraw certain permitted development rights. The proposals included a new 'Article 5A' which encompasses changes to doors, windows, roofs, painting or cladding, erection of porches and provision of hardstandings for cars. In addition, a proposed 'Article 5B' suggested a procedure to be followed when making an 'Article 5A direction'.

The government has also concluded that the new provisions should apply to front garden walls, fences and gates as the removal of these features has a significant effect on the character and appearance of conservation areas, including the detrimental visual impact of vehicles parking in front gardens. This would require amendment to the GDO and sections 2(1) and 3 of the Town and Country Planning (Demolition – Description of Buildings) Direction 1994. (This direction sets out descriptions of types of buildings, the demolition of which does not involve the 'development' of land.)

Following the end of the consultation period further progress on these proposals were announced in February 1995 (DoE, 1995) and the new provisions were given the relevant Parliamentary approval in June 1995 (see Appendix 6).

Other powers in conservation areas

Under Regulation 13, Schedule 3 of the 1990 Regulations, a number of provisions applicable to listed buildings may also be made applicable, with minor amendment, to unlisted buildings which are important to the character and appearance of conservation areas.

Urgent repairs notices

The urgent repairs procedure under s. 54 of the P(LBCA) Act 1990 may be applied to important unlisted buildings. This power has only been rarely used, but, with the development of CAPs, wider financing opportunities may result in more common use. For instance, the Grainger Town pilot proposals recognised the importance of unlisted historic buildings to the townscape of the area, many of which have been neglected and could be saved from destruction by greater use of this power. Moreover, the success of obtaining costs in contested cases will be very high, bearing in mind that the Secretary of State for National Heritage is required by s. 76 of the P(LBCA) Act 1990 to sanction the notice (Roberts and Sims, 1991; Smith, 1991).

Enforcement action

The regulations also refer to powers under ss. 38 and 42 of the P(LBCA) Act 1990. Under s. 38 an enforcement notice may be served if it is considered expedient, having regard to the effects of works (such as demolition without CAC) on the character and appearance of a conservation area. Under s. 42 it is possible for the LPA to execute and recover the cost of works required as specified in the enforcement notice if the appropriate steps have not been taken within the compliance period.

Appeals

Appeals in relation to issues concerning refusal of CAC, enforcement action as indicated above and against refusal of planning permission in a conservation area are dealt with in a similar manner, as is indicated in relation to listed buildings in Chapter 3.

Trees in conservation areas

It has been officially recognised that trees may play an important part in enhancing the appearance of a conservation area. Thus, trees which are not subject to a Tree Preservation Order (TPO) are subject to a provisional protection procedure under ss. 211–14 of the principal planning Act.

Accordingly, under s. 211 any person proposing to cut, top or lop, etc., a tree in a conservation area is required to give six weeks' prior notice to the LPA of the intended works. The purpose of this provision is to enable the LPA to make a TPO under s. 198 if it considers it necessary to do so on the grounds of amenity.

Failure to observe this provision is an offence. In *Bath City Council v. Pratt (T/A Cresent Investments)* (1987) unreported – Bristol Crown Court, the uprooting of a tree in the course of residential development works in a conservation area was found to be an offence despite the fact that the developer had argued that the tree had become dangerous. In fact the dangerous condition had only resulted from the development activity and had not existed prior to the works. It was held that the offence was one of strict liability and that the developer would be vicariously liable for the acts of his workmen. Penalties for contravention of the s. 211 procedure are similar to those for trees subject to a TPO and may include a requirement to replant (see DoE Circular 36/78).

However, there are two defences which may be raised against the charge that an offence has been committed: first, under s. 211(3)(a), that sufficient notice was given and the tree was properly identified to the LPA; secondly, under s. 211(3)(b)(i) that an act was carried out to a tree with the permission of the LPA, or s. 211(3)(b)(ii) following the six-week period but before the expiry of two years from the date the notice was served on the LPA. Yet failure to make a TPO within the six week period does not prevent the LPA from making one subsequently, as was established in the case of *R. v. North Hertfordshire District Council, ex parte Hyde* [1990] JPL 142.

There are certain exemptions from normal requirements contained in s. 211 which are provided by Regulation 3 of the Town and Country Planning (Tree Preservation Order) (Amendment) and (Trees in Conservation Areas) (Exempted Cases) Regulations 1975. These exemptions include the cutting down, uprooting, topping or lopping of trees which are dead or dying or have become dangerous, and various other miscellaneous exemptions. However, under s. 213 there is a requirement to replace a tree of an appropriate size and species at the same place as soon as is reasonably possible where trees in conservation areas are removed, uprooted or destroyed, unless the LPA dispenses with the requirements. Furthermore, the requirement to replace also applies to trees which are lawfully cut down because they are dead, dying or dangerous unless the LPA fail to make a TPO and the felling then occurs within the period provided by s. 211(3)(b)(ii) (Moore, 1995).

Under ss. 214A–D there are powers to apprehend an offence under s. 211 by way of an injunction, and provisions giving rights of entry on to land (with or without a warrant) to ascertain whether an offence has been committed.

Advertisement control

As indicated in Chapter 2, the Town and Country Planning (Control of Advertisements) Regulations 1992 allow for the designation of *Areas of Special Control* via a statutory power afforded by s. 221 of the principal planning Act. PPG 19 (Outside Advertisement Control) implies that the use of this procedure may be suitable to protect groups of buildings in sensitive locations, including in conservation areas. The procedure for defining such an area involves the LPA making an order which requires the Secretary of State's approval and is subject to an objection procedure under Regulation 5.

However, PPG 15 advises LPAs to be sensitive in the use of the regulations as outdoor advertising is essential to commercial activity 'in a free and diverse economy' (para. 4.32). It further advises that they may adopt policies for advertisement control as part of their duty under s. 71 of the P(LBCA) Act 1990 to formulate and publish proposals for the preservation and enhancement of conservation areas (para. 4.33). Thus a more cooperative stance is encouraged to educate against unsympathetic advertisements (para. 4.35). Moreover, English Heritage has also indicated the need for detailed guidance on advertisements, with particular reference to the importance of shop and trade signs being integrated into shop fronts or a building as a whole in sympathetic form, scale and materials to their context (English Heritage, 1993).

Yet LPAs have also been advised to seek greater control where other means have not sufficiently mitigated against the damaging effect of advertisements in conservation areas. The regulations are used generally to control advertisements on amenity grounds, taking into account the general characteristics of a locality, including the presence of any feature of historic, architectural, cultural or similar interest. Moreover, deemed consent to display specified classes of advertisements may be restricted in two ways. First, under Regulation 7, LPAs may approach the Secretary of State to obtain a direction to control most classes of advertisement, which may be for a specified period or indefinitely. For example, in some London Boroughs and in the City of Bath the Secretary of State has issued directions to control the use of estate agents' notice boards in conservation areas (Moore, 1994b). Secondly, under Regulation 8 a LPA may serve a notice requiring the discontinuance of the display of an advertisement to remedy a 'substantial injury to amenity' of an area, although this is subject to a right of appeal to the Secretary of State.

Regeneration projects in conservation areas

With the development plan system moving towards more environmentally led planning encompassing the concept of 'sustainable development', the process of finding conservation-based solutions to town regeneration initiatives is becoming ever more important. This approach has gained recognition from a number of sources.

The English Heritage guidance note on *Conservation Area Practice* highlighted the fact that some conservation areas have been in a state of relative decline and have suffered from lack of investment. In turn it recommended the development of economic regeneration strategies with the assistance of 'living over the shop initiatives' (LOTS), conservation area grants, city grants or the Economic Regional Development Fund (English Heritage, 1993). This builds on the pioneering work of the Civic Trust holistic initiatives such as at Wirksworth and Halifax and other local authority conservation–regeneration strategies such as the Jewellery Quarter (Birmingham), the Victorian Quarter (Leeds), the Lace Market (Nottingham) and Little Germany (Bradford), the benefits of which have now been recognised through the development of the CAP initiative (Burman, Pickard and Taylor, 1995).

Since 1994 further consideration has been given to the need for holistic regeneration through the DoE consultative document *Quality in the Environment*, the UK Strategy on *Sustainable Development*, and new funding arrangements provided by the *Heritage Lottery Fund, English Partnerships,* and the *Single Regeneration Budget.* Furthermore, the LOTS office has considered the possibility of setting up a specific BPT to deal with LOTS schemes on a nation-wide basis to take on buildings which are two daunting for local trusts (Petherick, 1995).

However, the main focus for conservation area regeneration strategies in the coming years will undoubtedly be the CAP initiative with joint funding arrangements between English Heritage and local authorities, supplemented by other partnership mechanisms between the public and private sectors. Moreover, English Heritage has indicated that these partnerships will replace current methods of funding in conservation areas (including 'town schemes') by April 1997, apart from a limited number of specific grants for certain outstanding historic buildings or particular buildings at risk (Johnson, 1994b).

The basis of a CAP is the development of a framework for concerted action by local authorities of using conservation-led solutions to resolve problems in some conservation areas presented by disrepair and dereliction through positive strategies for their preservation. Thus the idea is to identify problems and opportunities for channelling resources to encourage action on buildings, to develop management strategies, and to use development controls to protect the overall character of a conservation area (Johnson, 1994a). In order to take part in this initiative, LPAs have been invited to make a preliminary application. This must provide information on a range of matters:

- a brief assessment of the special architectural and historic interest of a conservation area as an initial justification for the provision of grant funds;
- a broad outline of the repair or enhancement needed in an area justifying the need for a grant-aided scheme;

- an outline of the extent, nature and cost of repairs anticipated by reference to basic conditions and repair problems;
- reference to necessary structural repairs, reinstatement works and proposals for environmental enhancement;
- a brief statement of existing conservation policies published in its local or development plan, or policies it proposes to set in place;
- information on the conservation experience, professional qualifications, and status of the person(s) who would run the partnership scheme for the local authority;
- the financial resources that may be made available by the local authority and the contributions that are expected from English Heritage.

Documentation about proposed schemes will be circulated each April, inviting bids for new CAP schemes to commence in the following April. During September each year English Heritage will come to a decision, in principle, on the preliminary applications submitted that year and will invite some authorities to prepare a more detailed action plan by the following January on the basis of a provisional indication of the scale of funding likely to be made available (English Heritage, 1994a).

The action plan sets out in more detail the proposed improvements to be achieved over a given time period and the depth and scale of resources required. In order to fully address the major and more complex problems associated with a conservation area, authorities have been asked to use a wide range of expertise to reflect the viewpoints of architects, designers, planners, engineers, valuers, historians and building professionals, including external expertise where the necessary professional skills are not available in-house. Moreover, a team approach has been advocated in order to produce a sustainable impact on an area (English Heritage, 1994b).

English Heritage has suggested a checklist of matters to which the action plan should refer:

1. *Assessment of special interest.* The nature of the architectural interest which justifies the designation of an area as a conservation area should be defined. In this respect LPAs have been directed to the assessment checklist provided in English Heritage's guidance note on *Conservation Area Practice* (see above).
2. *Defining the problems.* The analysis of problems faced by conservation areas may be the product of various factors. In this respect the analysis has been directed to consideration of the following issues:

- the manufacturing base of the area and relative employment statistics
- service or retailing in the area
- business confidence (where relevant)
- the area's economic base
- tourism potential
- housing and social mix and associated problems

- the area's identity and coherence.
- opportunities.

3. *Audit of the Fabric.* A detailed study is required on the issues that will need to be addressed in a partnership regeneration scheme. These include:

- major development proposals
- proposals and opportunities for environmental improvement
- an appraisal of the condition of the built fabric
- a survey of vacancy rates and under-use of upper floors
- opportunities for the enhancement of historic buildings
- consideration of traffic, car parking, pedestrian movement and opportunities for associated street improvements
- sample design briefs to indicate what could be achieved in relation to identified problem buildings
- definition of special action areas

4. *Programme and resources.* The action plan is required to indicate an implementation programme with a schedule of work targeting certain buildings. It must further indicate means to persuade private owners to participate, such as higher funding mechanisms than is normally given for repair work and associated estimates of grant aid needed. Further information has been provided to assist LPAs in assessing the eligible works for grant aid (English Heritage, 1994c) and specification requirements for supported works (English Heritage, 1994d) under the CAP initiative.
5. *Local authority conservation policies.* The action plan is also required to consider whether existing conservation area policies are adequate to safeguard the investment proposed in the partnership, particularly where key policies for an area and listed buildings contained within in it have not been put in place in the LPAs plan. Thus the action plan can be used as a preliminary to the adoption of new policies. Additional safeguards may also be needed in this context, such as an 'Article 4 direction'.
6. *Programme administration.* The action plan must set out the administrative arrangements proposed for running the partnership in terms of staff and accommodation requirements, responsibilities of the partners to the scheme, and management. The setting up of the administrative arrangements must also be indicated by an estimated timescale.

The final stage of a CAP initiative is the signing of a partnership agreement on the major issues. These include staffing, administration and accounting arrangements; the expected duration of the partnership, including the scale of resources to be provided from the partnership parties over the lifetime of the scheme; conditions relating to grant offers and workmanship; and provisions for monitoring and review.

The first 14 CAPs were launched in April 1994 from a shortlist of pilot projects volunteered. These reflect the potential for different types of partnership in the future. They involve at least one major urban centre (Newcastle upon Tyne, Liverpool), historic city (Lincoln), suburban area (Manningham area of Bradford), declining industrial centre (Baccup and Rawtenstall), coastal town (Scarborough, Hove, Hastings), market town (Wainfleet, Knaresborough), rural centre (Haltwistle, Wootton Bassett) spa town (Leamington), and the regeneration of a historic waterfront location (Greenwich) (Johnson, 1994a).

For example, the Newcastle upon Tyne scheme, involving the streets developed as part of a new centre by Richard Grainger in the nineteenth century and its surrounding area (collectively Grainger Town), has been chosen because of the scale of problems not resolved by previous measures. All 620 buildings in the area, both listed and unlisted, were surveyed in 1992. The result was that 50 per cent of listed buildings were found to be 'at risk' and a further 30 per cent were considered to be 'vulnerable', and 25 per cent of the unlisted buildings in the area were also found to be 'at risk' with 30 per cent 'vulnerable'. The need for special funding arrangements is only one of the problems which the scheme is now aiming to resolve (see Chapter 4). Moreover, most of the pilot schemes have chosen to adopt a higher grant-funding regime (between 40 and 80 per cent) than are provided by existing grant regimes such as through 'town schemes' or repair grants.

During 1994 all local authorities were invited to consider applying for the establishment of a CAP to run from April 1995. This resulted in 191 submissions from 128 separate local authorities, of which 115 proposals have been invited to develop action plans to take a share of £5.5 million funding allocated by English Heritage during 1995/96. Approximately half of the submissions were to convert existing 'town schemes' to CAPs while the remainder were for entirely new areas or areas which had not previously attracted a joint funding scheme between the local authority and English Heritage (Johnson, 1994b).

Undoubtedly the new partnership arrangements mark a shift in approach to the conservation of the built environment which builds on the government's policy desires for 'sustainability'. However, it will take perhaps until the end of this millennium before the effectiveness of these new arrangements can properly be ascertained. Yet the evidence of early examples of holistic regeneration associated with the historic environment, involving a partnership of public and private interests, indicates the considerable potential of this approach as compared to dealing with issues in isolation.

Case study: Seven Dials, Bath

The development of any site in the centre of the city of Bath is particularly sensitive, bearing in mind its unique townscape of Georgian architecture which has been designated as a 'World Heritage Site'. The following case study provides a good exemplar for the approach to be adopted by developers in terms of satisfying the requirements of s. 72 to *preserve* and *enhance* the *character* and *appearance* of a conservation area by the provision of new development which harmonises with the existing fabric, and provides new life to the area as well as appropriate environmental improvements.

The Seven Dials site had previously contained a number of eighteenth-century buildings which undoubtedly would have been listed had they survived to the present day. Due to bomb damage during the Second World War these buildings were demolished and the site cleared. For some time afterwards the site was loosely used for vehicle parking. However, in the 1960s the site was redeveloped with a one-storey nightclub and roof garden centre. It is evident that this development was out of scale and did not blend with the character of the three- and four-storey historic buildings surrounding the site, but at this time conservation controls were not as strong as today. Although the city of Bath was one of first areas in the country to designate a conservation area, this was not until 1969.

The developer's general approach to new development schemes in historic towns was to investigate the historical development and materials used in an area and judge how a new development proposal could be integrated within the existing fabric (Green, 1992). The city of Bath presented a particular problem in terms of integrating new development as most of the historic city as existing was developed over a period of approximately 70 years, using the indigenous Bath stone in a predominantly classical style. Moreover, the LPA had placed an embargo on new office development to restrict the impact of traffic in the historic environment which would only be waived for a local user or where the environment would be enhanced (Pettifer, 1992). Furthermore, the site itself added further problems being triangular in shape and sloping, with adjoining outstanding historic buildings, including the former eighteenth-century Baroque townhouse of Beau Nash and Theatre Royal, and was also scheduled as an ancient monument (see Chapter 8).

In order to resolve potential resistance to a new scheme the developer sought specialist advice by commissioning a report from the Civic Design Partnership which recommended a scheme which would seek to provide sympathetic new development, respecting the integrity of the surrounding townscape and historic spaces, based on painstaking architectural and historical analysis regarding the site and its surroundings (Davies, 1991). The LPA was at first reluctant to see any replacement development greater in height than the 1960s one-storey nightclub development; but with some foresight the developer saw the opportunity for blending new development with the existing historic fabric on a larger scale. Research of the history of the site revealed photographs of three- and four-storey Georgian buildings which had been demolished due to war damage. These were used to provide an argument for allowing a scheme on a larger scale and overturn the LPA's desire for a single-storey replacement (Figs 7.6, 7.7 and 7.8).

A scheme was formulated incorporating a mixture of retail, office and restaurant uses designed by in-house and consultant architects and the first drawings of the proposed development were completed in 1988. A classical

Fig. 7.6 Seven Dials site, Bath, in 1936, showing four-storey buildings which subsequently had to be demolished due to war damage. *(Reproduced with the permission of Avon County Library (Bath Central Library).)*

design was chosen but one which would represent the modern age with less detailing compared to its historical neighbours. The developer had argued that this approach was necessary if the past was not to be falsified. Extensive discussions were carried out between the developer's design team, architectural consultants and the LPA's historic building officers. The final design scheme arose from a collaboration of ideas over a two-year period. This time was required to resolve the design issues and other archaeological factors encountered with the site (see Chapter 8).

The final design incorporated both Regency and Georgian elements, but with detail kept to a minimum. This included a rotunda in Regency style

Fig. 7.7 Seven Dials site, Bath, in 1960, when used as a car park. (*Reproduced with the permission of Avon County Library* (*Bath Central Library*).)

Fig. 7.8 Seven Dials site, Bath, in 1974, with a single-storey nightclub development. (*Courtesy of The Bath Chronicle.*)

incorporating wrought iron railings with decorative leadwork on the balcony canopy and dome above, and a colonnaded frontage (Fig. 7.9). Local ashlar stone blocks were used in the façade, and although this was lighter in shade as compared to neighbouring buildings, it was estimated that it would start to blend in with surrounding buildings after approximately five years. The scheme was set back so that it would not visually encroach on a neighbouring Baroque frontage (Fig. 7.10).

By far the greatest design problem to surmount was the sloping site. The use of steps on the retail side to the scheme was rejected on the grounds that it would not be conducive to attracting a pedestrian flow. This was resolved by retaining an even slope to the pavement while using a curved entablature to support the column design (Fig. 7.11), derived from a precedent found in a terrace of houses elsewhere in the city (Fig. 7.12), and a horizontal and curved parapet to the top of the building.

The end result has been a successful development in investment terms, as evidenced by the fact that the developer's 150-year leasehold interest was sold in 1994 for nearly £6 million (Pearce, 1994). Significantly it has also made a conservation 'gain' to the city of Bath. This was motivated by a development company which has built up considerable experience of working in historic centres through creating designs sensitive to local environments. The development has clearly assisted in the preservation and enhancement of the character and appearance of the Bath city conservation area by replacing a building which detracted from its surroundings with one which complemented neighbouring buildings. Moreover, the scheme has resulted in substantial streetscape improvements without creating additional traffic penetration into this area of the city and has brought new business to a relatively dead corner. There was considerable imagination in the design work, not just in the curved entablature and sloping colonnade on the retail pitch but also in creating a courtyard space in the interior with a fountain feature (Fig. 7.13), by hiding an electricity substation within the building, utilising the roof space to screen service plant and providing ventilation through a carefully designed advertising bollard located in the pavement, which is now used by the adjoining Theatre Royal.

The Seven Dials success is largely due to the approach of the developer not only in having a vision for an opportunity site but in persuading the LPA of the merits of this vision by extensive historical research of the site and its surrounding environment and then by successfully cooperating with the LPA in design discussions. This emphasises the importance of conservation area appraisal and pre-application discussions. The scheme was commended by H.R.H. the Prince of Wales, awarded a design award by the Royal Institute of British Architects and was the subject of an exhibition with other commended design schemes entitled *Before and After Planning* at their headquarters in January 1993, and it also received favourable attention from English Heritage. Of significance, the developent is a new design and while it has incorporated classical interpretations, which was favoured in this particular conservation

Fig. 7.9 Seven Dials site, Bath, as redeveloped in 1992, showing the Regency rotunda, railings and dome, and colonnaded frontage.

Fig. 7.10 Seven Dials site, Bath, as redeveloped in 1992, showing street improvements with ventilation plant hidden in the advertising bollard for the adjoining New Theatre Royal, formerly part of Beau Nash's house.

Fig. 7.11 Column design with curved entablature.

area, it has not slavishly copied other buildings in the area. But the new building alone is not the only important factor which has assisted in enhancing the conservation area as the scheme brought road and pavements which have, in turn, improved the setting of other historic buildings.

Elsewhere it may be more possible to create truly modern designs for new buidings which will enhance the character of a conservation area using materials and a building scale which are in sympathy with the character of an

Fig. 7.12 Bathwick Place. The precedent used for the curved entablature.

Fig. 7.13 Seven Dials site, Bath, as redeveloped in 1992, with interior courtyard space in natural stone paving and fountain feature.

area. But in the centre of Bath it was considered necessary to pay homage to the Georgian atmosphere with a modern interpretation.

References

Andreae, S. (1992) Conservation areas in London. Paper presented at the London Conservation Areas Conference, held at the Art Workers' Guild, London, 25 November, pp. 2–5.

Bloxsidge, R. (1975) The local authority contribution, *Town Planning Review*, Vol. 46, No. 4, pp. 466–9 (Special edition to commemorate European Architectural Year 1975).

Booth, E. (1994) Paving: the good, the bad and the ugly, *Context*, No. 41, March, pp. 10–11.

Burke, G. (1976) *Townscapes*, Pelican Books, pp. 116–40.

Burman, P., Pickard, R.D. and Taylor, S. (ed.). (1995) Papers presented at a consultation entitled *The Economics of Architectural Conservation*, held at the Institute of Advanced Architectural Studies, University of York in association with the European Union Department DG–10, 12–14 February.

Davies, P. (1991): 'Improvements' in historic areas, *Conservation Bulletin*, Issue 15, October, pp. 15–16.

Davies, P. (1992) Conservation Areas, *Conservation Bulletin*, Issue 17, June, pp. 1–2.

DNH (1994) Department of National Heritage Press Release No. 54/1994.

DoE (1995) Department of the Environment Press Release No. 94/1995.

English Heritage (1993) *Conservation Area Practice: English Heritage Guidance on the Management of Conservation Areas*.

English Heritage (1994a) *What is a Conservation Area Partnership?*, English Heritage conservation area partnership schemes, Leaflet 1.

English Heritage (1994b) *Preparing the Action Plan*, English Heritage conservation area partnership schemes, Leaflet 2.

English Heritage (1994c) *A Guide to Eligible Work*, English Heritage conservation area partnership schemes, Leaflet 3.

English Heritage (1994d) *Specification Requirements for Grant-aided Work*, English Heritage conservation area partnership schemes, Leaflet 4.

EHTF (1992): *Townscape in Trouble: Conservation Areas – the Case for Change*, English Historic Towns Forum.

Graves, P. and Ross, S. (1991) Conservation areas: a presumption to conserve, *The Estates Gazette*, 21 September, pp. 108–10.

Green, N. (1994) Protecting the street scene: traditional paving, *Context*, No. 41, March, pp. 7–8.

Green, S. (1992) Planning implications of developing in conservation areas in historic cities. Presentation at the RICS Surveyors Briefing 1992: Annual Conference, Bath, 25–6 September.

Grover, P. and Lewis, R. (1994) A paving strategy, *Context*, No. 41, March, pp. 12–13.

Harris, R. (1994): Streetlighting in village conservation areas – where now?, *Context*, No. 41, March, p. 8.

Johnson, S. (1994a) Increasing local conservation commitment through partnership schemes, *Conservation Bulletin*, Issue 23, July, pp. 26–8.

Johnson, S. (1994b) Conservation area partnerships, *Conservation Bulletin*, Issue 24, November, pp. 6–8.

Lawton, R. (1994) Paving: becalmed in Barton upon Humber, *Context*, No. 41, March, pp. 14–15.

Lillford, R. and Dadson, P. (1994) Preservation and enhancement of historic areas: the Somerset approach, *Context*, No. 42, June, pp. 29–31.

MacQueen, R. (1992) Permitted development in conservation areas. Paper presented at the London Conservation Areas Conference, held at the Art Workers' Guild, London, 25 November, pp. 6–11.

Martin, J. (1995) Development in conservation areas – the challenge, *The Estates Times*, 4 February, pp. 156–7.

Millichap D. (1990) Conservation areas – Steinberg and after [1990] JPL 233.

Millichap, D. and Judd, M. (1991) *Townscape in Trouble Seminar Papers*, English Historic Towns Forum.

Moore, V. (1994a) *A Practical Guide to Planning Law*, 4th edn, Blackstone Press, pp. 374–79.

Moore, V. (ed.) (1994b) Practical points: Duties of planning authorities in respect of conservation Areas [1994] JPL 298.

Moore, V. (ed.) (1995) Practical points: Control over felling of trees in conservation areas [1995] JPL 273.

Moreton, D. (1991) Conservation areas – has saturation point been reached? *The Planner*, 17 May, pp. 5–8.

Moreton, D. (1994) Sir Titus, *Context*, No. 41, March, pp. 28–9.

Mynors, C. (1993a) M'Learned Friend, *Context*, No. 37, March, pp. 27–8.

Mynors, C. (1993b) M'Learned Friend, *Context*, No. 40, December, pp. 23–4.

Pearce, H. (ed.) (1994) News: Chartwell sells Seven Dials, *The Estates Gazette*, 22 January, p. 34.

Pearce, G., Hems, L. and Hennessy, B. (1990) *The Conservation Areas of England*, English Heritage.

Petherick, A. (1995) Living over the shop – 'the successes'. A seminar held at the Institute of Advanced Architectural Studies, University of York, 16 March.

Pettifer, M. (1992) Sense and sensitivity, *New Builder*, No. 137, 9 July, p. 21.

Purdue, M. (1994) The impact of section 54A [1994] JPL 399.

Ross, M. (1991) *Planning and the Heritage: Policy and Procedures*, E. & F.N. Spon, pp. 28–36 and 118–37.

Reynolds, J.P. (1975) Heritage Year in Britain, *Town Planning Review*, Vol. 46, No. 4, pp. 355–64. (Special edition to commemorate European Architectural Year 1975.)

Roberts, G. and Sims, A. (1991) Urgent Repairs Notices, *Conservation Bulletin*, Issue 15, October, p. 20.

Smith, J. (1991) Section 54 notices and non-listed buildings, *Context*, No. 29, March, p. 35.

Stubbs, M. and Lavers, A. (1991) Steinberg and after: Decision making and development control in conservation areas [1991] JPL 9.

Chapter 8

ANCIENT MONUMENTS AND THE ARCHAEOLOGICAL HERITAGE

Introduction

Following the formation of SPAB in 1877, the development of the conservation movement in the late nineteenth century led to the first legislative provisions for the protection of historic items. The Ancient Monuments Act 1882 gave legal protection to 29 monuments in England which were set out in a schedule to the Act, hence the term *Scheduled Ancient Monument* which survives today. The ancient monument legislation predates that for listed buildings as the concept of listing was first introduced by the Town and Country Planning Act 1944 (Ross, 1991).

The provisions relating to ancient monuments were strengthened by the subsequent Ancient Monuments Protection Act 1913 and by the Ancient Monument Consolidation and Amendment Act 1931. These have since been repealed and their provisions consolidated in the Ancient Monuments and Archaeological Areas Act 1979 (the 1979 Act) which also introduced the concept of *Areas of Archaeological Importance.*

The Ancient Monuments Board, established by the 1913 Act, was set up to act as an expert committee to deal with the care of monuments until 1984. Following the National Heritage Act 1983 this responsibility was given to the new Historic Buildings and Monuments Commission for England (English Heritage) which acts as the Secretary of State's official adviser on matters relating to ancient monuments. Specific functions in this respect include the provision of advice on whether particular remains are worthy of scheduling and on applications for scheduled monument consent (SMC) regarding works to protected monuments; the provision of financial assistance towards the upkeep of ancient monuments and towards archaeological investigation; and, with the consent of the Secretary of State, the acquisition or guardianship of ancient monuments. Since the formation of the DNH in 1992 casework responsibilities for the scheduling of ancient monuments and for granting SMC were passed to the new government department from the DoE. The DoE nevertheless retains an interest in planning issues where, for instance, a site which includes archaeological remains (whether scheduled or not) is the subject of a development proposal.

Some buildings are both scheduled as ancient monuments and listed for their special architectural or historic interest. The overlap is indicative of the fact that scheduling predates listing. Where both protections apply, SMC takes precedence over LBC. Moreover, scheduling provides a stricter control concerning repair work. However, para. 6.34 of PPG 15 indicates that, for the future, the type of protection which is most appropriate in the circumstances will be applied. This is being addressed in the English Heritage's *Monuments Protection Programme* (MPP), designed to update information on ancient sites and identify further sites and monuments that are suitable for scheduling, which has been planned to run until 2007. Paragraph 4.25 of PPG 15 also clarifies that SMC takes precedence over conservation area control where protected monuments are located within conservation areas.

Since the late 1980s, when a number of significant archaeological discoveries were made during the course of new development activity, government policy regarding ancient monuments and archaeological remains has been strengthened. In 1990 Planning Policy Guidance Note 16: Archaeology and Planning (PPG 16) was issued to provide guidance for LPAs, property owners, developers, archaeologists, amenity societies and the general public with particular reference to the handling of archaeological remains in the development plan and development control systems. Paragraphs 15 and 16 of PPG 16 specifically indicate that development plans should include policies for the protection, enhancement and preservation of sites of archaeological interest and their settings including scheduled sites and other unscheduled archaeological remains of more local importance. Paragraph 18 further indicates that the desirability of preserving an ancient monument and its setting is a material consideration in determining a planning application whether the monument is scheduled or not. This follows the decision in *Hoveringham Gravels v. Secretary of State for the Environment* [1975] 2 WLR 897.

Since the introduction of the presumption in favour of the development plan for the determination of planning applications under s. 54A of the Town and Country Planning Act 1990, greater weight has been placed on the inclusion of appropriate policies to recognise the value, variety and vulnerability of England's archaeological remains. In 1992 English Heritage published an advice note entitled *Development Plan Policies for Archaeology* to assist LPAs in the formulation of appropriate policies in this respect (English Heritage, 1992).

Preserving the heritage of ancient sites, buildings and remains has been given a high priority within the wider historic environment. Moreover, a European framework for the guidance contained in PPG 16 has been provided in a revised *European Convention on the Protection of the Archaeological Heritage* which was signed by the government in January 1992 (Wainwright, 1992) and on a international level by the ICOMOS *Charter for the Protection and Management of the Archaeological Heritage*, commonly known as the *Lausanne Charter*, in 1990 (see Appendix 5). Thus, in turn, the preservation of the archaeological heritage has an impact upon owners of both protected and unprotected sites and upon anyone proposing activities which may cause harm to such sites.

Scheduled Ancient Monuments

The process of protecting ancient monuments

Under s. 1 of the 1979 Act the Secretary of State is under a duty to compile and maintain a schedule of ancient monuments of national importance. Anyone may put forward a monument for scheduling, though most proposals originate from English Heritage, and in any event the Secretary of State is required to consult English Heritage before adding to the schedule or in the rare cases where monuments are to be descheduled. There is no statutory requirement for owners to be consulted before a monument is scheduled although owners are normally consulted, and, once scheduled, details of a monument are published by English Heritage on a county by county basis and scheduled sites are registered as charge in the local registry and notified to the particular county archaeologist (or English Heritage in London) for inclusion in the local sites and monuments records (SMR).

For the purposes of determining what may become subject to the scheduling procedure, s. 67 of the 1979 Act defines a *monument* as being *any building, structure or work above or below ground; any cave or excavation; any site comprising the remains of any such building, structure or work, or of any cave or excavation; and any site comprising the remains of a vehicle, vessel, aircraft or movable structure or part thereof.* The definition specifically excludes an ecclesiastical building used for ecclesiastical purposes (which may be protected by listing) and a wreck or the site of a wreck which has been protected by an order under the Protection of Wrecks Act 1973. It is also not possible to schedule an occupied dwellinghouse other than where the occupation is by a person employed as a caretaker there.

Section 67 also defines an *ancient monument* as a monument which, in the opinion of the Secretary of State, is of public interest by way of any *historic, architectural, traditional, artistic, or archaeological* qualities attached to it. The scheduling procedure therefore encompasses a wider interest consideration than the *special architectural or historic* interest required for listing purposes and also may be distinguished by the fact that the site of the particular item, as well as the item itself, may be protected. Once an ancient monument is scheduled it is known as a Scheduled Ancient Monument although the term *ancient* is something of a misnomer as there is no date criteria for scheduling. In fact there are structures dating from the twentieth century which have been scheduled. Nevertheless the majority of monuments are in fact ancient, as was evidenced by a 1984 report entitled *England's Archaeological Resource* which identified that 61 per cent of the total number of ancient monuments were prehistoric and that only 7 per cent were post-medieval (IAM, 1984).

The decision as to whether a monument should be scheduled is purely at the discretion of the Secretary of State. In *R. v. Secretary of State for the Environment, ex parte Rose Theatre Trust Co.* [1989] QBD 257, Schieman J. decided that the Secretary of State had a broad discretion but must not take anything legally irrelevant into account when deciding whether a monument, which appeared to him to be of national importance, should be scheduled.

The case arose out of a campaign to preserve the remains of the Shakespearian Rose Theatre discovered in 1989 during the course of new development works. However, the campaign to have the remains protected by scheduling was defeated as the judgment found that the Secretary of State had properly exercised his discretion in relation to specific points of challenge. First, he was entitled to have regard to the relevant risk of a claim for compensation for revocation of planning permission and loss of development value, estimated to be in the region of £100 million (Johnston, 1989). Secondly, he was not precluded from considering at the scheduling stage the need to balance the competing interests of preservation against the need for new development. Thirdly, the desirability of achieving voluntary cooperation with the developer was a proper matter for consideration, particularly if this would result in the site being protected by other means. In fact the developers had agreed to fund an archaeological investigation of the site which revealed the remains and agreed to further delay after the Secretary of State offered £1 million towards the cost implications. Subsequently, the developers agreed to build their office block on stilts over the site with the remains of the theatre being preserved and displayed although not scheduled, the eventual cost of delays and alterations to the building plans being estimated to be at least £10 million to the developers (Redman, 1990).

The only statutory criterion which the Secretary of State must consider in exercising his discretion is that a monument must appear to be of national importance. In 1983 the DoE published non-statutory guidelines upon which the assessment of national importance is to be judged. Although not to be regarded as definitive, the criteria are indicators which contribute to the wider judgment to be used in individual cases. The criteria have since been reproduced in Annex 4 of PPG 16 and include all of the following indicators:

- *Period.* All types of monument that categorise a category or period should be considered for preservation.
- *Rarity.* Some monument categories are so scarce that all surviving examples should be preserved notwithstanding the need to preserve examples which are commonplace.
- *Documentation.* The existence of records and other supporting evidence may enhance the significance of a monument.
- *Group value.* The value of a monument may be enhanced by its association with a number of other monuments such as the association of a field system with a settlement and a cemetery.

- *Survival/Condition.* The extent of original fabric remaining and its condition are relevant in determining whether a monument should be protected.
- *Fragility/Vulnerability.* This criterion has particular relevance to field monuments which are vulnerable to damage from a single ploughing operation or other unsympathetic treatment.
- *Diversity.* Some monuments may have a combination of factors which make them suitable for scheduling, while others may be protected because of a single important attribute.
- *Potential.* Where evidence of remains are not conclusive, and could only be confirmed by thorough excavation, a site may nevertheless be protected on the grounds of anticipation of existence and importance.

Shortly after English Heritage was formed in 1984 an initiative was devised to deal with the inadequacies of the existing schedule revealed by the report on *England's Archaeological Resource.* In fact it was estimated that there were then over 300,000 recorded sites and finds in England and that the eventual number of records could be over 600,000. To form a reasonable sample it was considered that some 60,000 monuments should be protected, substantially increasing on the then existing 13,000 scheduled sites (Startin, 1988). However, the intended *Scheduling Enhancement Programme* proved difficult to commence largely due to the government being unable to fund the work. Furthermore certain preparatory work was necessary including the design of a computer system to support scheduling procedures and to maintain records in a way that would allow easy interchange of information between the new record and the existing National Archaeological Record held by the RCHME. The original concept of the initiative was subsequently revised and expanded to include a review of those monuments already scheduled and then to pursue greater knowledge of other aspects of England's archaeological resource, including making further scheduling recommendations and identifying and recording items where scheduling would not be appropriate. This enhanced initiative was entitled the *Monuments Protection Programme* (MPP).

The limitations of the original initiative led to only about 50 new schedulings a year. However, under the MPP 400 recommendations were made in 1990/91 and 1240 in 1993/94. In 1994 it was estimated that there will be over 30,000 scheduled ancient monuments covering some 50,000 individual archaeological sites by the time MPP has run its course (Wainwright, 1994).

The consequences of scheduling

There is no right of appeal against a monument being included in the schedule although an owner may apply to the Secretary of State for exclusion or amendment for which there is a power to make an order to this effect under s. 1 of the 1979 Act. If an order is not confirmed, SMC must be obtained before works can be undertaken which may affect a scheduled monument.

Under s. 2 of the 1979 Act any person who carries out unauthorised works to a scheduled monument will be guilty of an offence. The unauthorised activity includes works which result in demolition, destruction or damage; or works of removal, repair, alteration or addition; or flooding or tipping operations in, on, or under the land where a scheduled monument is located. The effects of scheduling are therefore more severe than the listing procedure, particularly as repair work requires offical consent. Under s. 28(1) of the 1979 Act there is an offence of destroying a *protected monument* (meaning a scheduled monument which is under the ownership or guardianship of the Secretary of State, English Heritage or a local authority), which is normally used to prosecute cases of vandalism. The use of a metal dectector, which may cause damage not only to the fabric of a scheduled monument but also to its interpretation and understanding if artefacts are removed, is an offence under s. 42(5).

The penalty for failing to obtain SMC, or failing to comply with a condition attached to SMC, or in relation to s. 28(1), or under s. 42(5) for removing artefacts without consent, is a fine up to a statutory maximum of £5000 in a Magistrates' Court, or an unlimited amount on conviction on indictment at the Crown Court. The use of the s. 28(1) procedure has a further implication of a possible term of imprisonment for a period not exceeding two years. The use of a metal detector without consent, but without removing artefacts, is subject to a fine of £200. The maximum level of fine on summary conviction is updated from time to time in line with inflation. However, in 1994 English Heritage proposed an increase of the maximum fine level to £20,000, which would provide unity with the listed building fine level (see Chapter 2), but this will require an amendment to the 1979 Act (Brainsby, 1994).

English Heritage prepared a guidance note on action following an offence entitled *Damage to Ancient Monuments: Guidance to Prosecutions* which was issued by the Association of County Councils in 1988. Some local authorities have instituted prosecution proceedings, the power to do so not being limited to the Secretary of State or English Heritage. However, English Heritage tend to take the main role in commencing proceedings by keeping a record of reported incidents affecting scheduled monuments and frequently carrying out preliminary investigations with police assistance before the Crown Prosecution Service is approached to institute proceedings.

For example, the eighth Marquis of Hereford was fined £10,000 on indictment for ploughing a field which resulted in serious damage to Roman remains. On appeal to the Court of Appeal (*R. v. Seymour* [1988] 1 PLR 19) the fine was reduced to £3,000, the court taking the view that the original fine would have been more appropriate to a question of flagrant disregard of a monument for the purposes of personal gain, rather than the negligence or inadvertence involved in this case. A number of other cases have been publicised by English Heritage which reveal that the Crown Court and Court of Appeal are willing to impose substantial fines for unauthorised work. In particular, in the 1992 case of *R. v. J.O. Sims Ltd* (unreported), having previously pleaded guilty to one charge of causing or permitting unauthorised works to the scheduled site of the former Winchester Palace and associated Roman remains, a fine of £75,000, with £1000 costs, was imposed for removing major stone and chalk walls from the site (Carter, 1993).

There are four defences to the s. 2 provisions contained in ss. 2(6)–2(9). In broad terms these include defences of due diligence, i.e. that all reasonable precautions were taken to avoid contravening the conditions of a SMC; or to avoid destruction or damage destruction; or, in cases of concealed monuments, that the defendant did not know, or had no reason to believe, that the monument was in the area affected by the works. Moreover, in *R. v. Jackson* [1994] (*The Independent*, 23 May 1994) the Court of Appeal confirmed that the precise extent of a scheduled monument should be capable of being ascertained from relevant material documents relating to it, which should ordinarily include the notification of scheduling and an accompanying map. This case follows a Crown Court case concerning alleged damage to a scheduled monument, Condicote Henge, in Gloucestershire, which was dismissed on the grounds that evidence of the existence of the monument was not made available (Fairclough, 1989). Thus criminal liability under s. 2 cannot depend on speculation. A further defence is allowed on the basis of works being urgently necessary in the interests of health and safety, similar to the provisions available in the case of listed buildings which have become *dangerous structures*, although the defence is less restrictive only requiring notice in writing to the Secretary of State as soon as is practicable. In general terms these defences are drafted less strictly than the defences relating to listed buildings where the offences are absolute, i.e. breaches of LBC constitute offences of strict liability (Suddards, 1988).

Scheduled monument consent

Works to a scheduled monument are authorised under s. 2(3) of the 1979 Act by the granting of SMC, which may only be granted by the Secretary of State after consulting English Heritage, so long as they are carried out in accordance with the terms of the consent and any conditions which may be attached to it under s. 2(4). There is no provision for granting outline consent but there are ten *class consents* which enable owners to proceed with certain specified types of work without formally applying for consent (see below). Under s. 50 of the 1979 Act monuments on Crown land may be scheduled, but until the proposed removal of Crown Exemption (see Chapter 2) there is an exemption from statutory SMC controls, though a non-statutory procedure known as *scheduled monument clearance* applies instead.

The procedures for dealing with applications for SMC are contained in the Ancient Monuments (Applications for Scheduled Monument Consent) Regulations 1981. Guidance notes, obtainable from the DNH or English Heritage, further explain how to apply for SMC. In particular, prospective applicants are advised to discuss their proposals with English Heritage at the earliest opportunity.

The basic requirements for submitting an application for SMC under the regulations are that the prescribed form of application must be accompanied by plans and drawings relevant to the works and the site, an appropriate certificate of ownership and any other information which the Secretary of State requests. The procedure for notification of owners is similar to the procedure in the case of LBC with 21 days' notice being required before an application is submitted. There is no requirement to advertise applications for SMC. Applicants are normally notified of the provisional views of the Secretary of State before a decision is formally issued. The reason for this is largely due to the fact that, before determining the application, the Secretary of State is required by Schedule 1 of the 1979 Act to either give the applicant (or any other person to whom it appears expedient) the opportunity of making representations at an informal hearing or hold a public local inquiry. Where a hearing or inquiry is to be held for a proposal which is also the subject of a planning inquiry, it is generally considered to be expedient to hold the two inquiries at the same time. In other respects there is no requirement to involve a LPA in matters relating to a SMC application.

The majority of reported decisions on applications for SMC are concerned with proposed works which constitute development and require planning permission. Paragraph 8 of PPG 16 indicates that *'where nationally important archaeological remains ... are affected by proposed development there should be a presumption in favour of preservation'*. However, an unusual case illustrates that, on occasion, the Secretary of State may exercise his discretion in a different way. The case involved an application concerning a proposal to construct a radio station and mast on the scheduled site of the Trundle, near Chichester. The Secretary of State determined the application on the basis of weighing harm to the cultural heritage against telecommunication needs (the subject of separate planning guidance). The view was taken that the proper test to be applied was not limited to the effect the proposed development would have on the monument both visually and physically in terms of its effect on the archaeological remains, but should include other matters of public interest, in this case the need to erect a mast on the site (Moore, 1991).

There is no right of appeal against a decision to refuse SMC except by judicial review in the courts on a point of law. However, under s. 7 of the 1979 Act compensation may be payable for the refusal of SMC. Four requirements must be satisfied before compensation will be awarded. These are:

- that the claimant must have an interest in the whole or any part of a scheduled monument;

- that expenditure has been incurred or the claimant has otherwise suffered loss or damage;
- that this has occurred as a consequence of the refusal or the granting of SMC subject to conditions;
- that the claim relates to certain specified works.

The works include:

- those reasonably necessary for carrying out development authorised by a planning permission (which is still valid) before the monument was scheduled;
- works not constituting development or authorised by the Town and Country Planning General Development Order 1988; works reasonably necessary for the continuation of any use of a monument which existed immediately prior to the date of application for SMC (see Appendix 7).

The scope for claimimg compensation for not being able to carry out these works is further limited by conditions provided in s. 7(3)–(5). All claims for compensation are paid by English Heritage. Under s. 8 there is a provision for the recovery of compensation if SMC is subsequently granted or if conditional SMC is modified so that the matters which attracted the claim for compensation cease to apply.

Where SMC is granted it is subject to s. 4 of the 1979 Act which provides a time limit on the commencement of consented works of five years, or of a shorter period if specified. Moreover, as well as varying the period of consent the Secretary of State may, after consulting English Heritage, make a direction to modify or revoke a consent to any extent which he considers expedient. A claim for compensation for abortive expenditure may be payable in this instance.

Class consents

An owner may carry out certain specified works where there is no undue risk of damage to a scheduled monument without the need for an application for SMC. The Ancient Monuments (Class Consents) Order 1981, which was amended in 1984, provided for six classes of works for which SMC was deemed to be granted. In 1993 the DNH announced the need to update certain aspects of the then existing Class Consent Orders via a consultation paper. The subsequent Ancient Monuments (Class Consent) Order 1994, revoking previous orders, provided revisions to existing class consents 1, 2 and 5 and provided four new class consents. The work for which consent is granted now comprise the following:

1. *Agriculture, horticulture and forestry works.* This includes certain of the above works as the same kind as has been carried out lawfully in the same location and on the same spot within that location during the

previous six years. Certain types of work are specifically not permitted by the order. These include ploughing at a depth below that which has been previously carried out legally; disturbance of the land below the depth of 300 millimetres; subsoil or topsoil works in connection with drainage, planting or uprooting, tipping or for obtaining turf; disturbance of any building, structure, work or remains; the erection of any building or structure; or the laying of paths, hard-standings, foundations or the erection of fences or barriers other than where the works are in connection with domestic garden works.

2. *Works by British Coal Corporation or their licensees.* This includes works executed more than 10 metres below ground by the Corporation or their licensees.
3. *Works by British Waterways Board.* This includes works required to be undertaken by the Board for the purposes of ensuring the function of a canal which do not involve any material alterations to a scheduled monument.
4. *Works for the repair or maintenance of machinery.* This includes works for the above which do not involve any material alterations to a scheduled monument.
5. *Works urgently necessary for health and safety.* This includes relevant works provided that they are limited to the minimum measures necessary and that they are justified by notice in writing to the Secretary of State at the earliest possible opportunity.
6. *Works by the Commission.* Any works required to be undertaken by English Heritage are permitted.
7. *Works of archaeological evaluation.* Such works carried out by or on behalf of a person who has applied for consent are permitted provided they are: for the purposes of providing information required by the Secretary of State for the determination of the application; carried out under the supervision of a person approved by the Secretary of State; and in accordance with a written specification approved by the Secretary of State or English Heritage.
8. *Works carried out under certain agreements concerning ancient monuments.* Works permitted under this class include works of preservation or maintenance executed in accordance with a written agreement between the occupier of a monument and its site and the Secretary of State or English Heritage under s. 17 of the 1979 Act.
9. *Works grant aided under section 24 of the Act.* Section 24 of the 1979 Act allows English Heritage to defray or contribute to the cost of works to a monument. Such works must be carried out within the terms of a written agreement between the occupier of the monument and the Secretary of State or English Heritage.
10. *Works undertaken by the Royal Commission on the Historical Monuments of England.* This involves works for survey purposes to a depth not exceeding 300 millimetres carried out by the RCHME.

Monument Management

Various procedures are provided by the 1979 Act to assist in the future preservation of a scheduled monument.

Under ss. 10 and 11 of the 1979 Act the Secretary of State may acquire a scheduled monument in order to secure its preservation by way of a gift, by agreement or in limited cases by compulsory acquisition. Section 15 allows other adjoining land to be compulsorily acquired if it appears reasonably necessary to take such action for maintenance, management and public access purposes. However, the more common way of securing preservation is by taking a scheduled monument (and adjoining land if necessary) into guardianship under provisions contained in ss. 12–15. These are voluntary arrangements which result in the guardian (the Secretary of State, English Heritage or a local authority) taking full responsibility for the care, maintenance and management of a monument, with the owner normally being required to permit public access.

Separate to the guardianship procedure, an occupier of a scheduled monument may enter into a management agreement (with the Secretary of State, English Heritage or a local authority) under s. 17 concerning the monument itself and adjoining land. An agreement may make provisions for such matters as the maintenance, preservation and other works to be carried out; public access; restrictions on use; and the making of payments to the occupier. Moreover, they often run for an agreed period of time in return for a lump sum payment to ensure a beneficial management. Such agreements are most frequently used in connection with field monuments on agricultural land generally. Section 17 agreements will be binding on successors in title to the original party.

In other respects the responsibility for maintaining a scheduled monument remains with the owner. The owner may be assisted by grant aid provided by English Heritage under s. 24 of the 1979 Act principally for the costs of repair, archaeological recording or the consolidation of a monument although this is limited by public finance constraints in the same way as in the case of listed buildings. Grant awards may occasionally be given towards the cost of presenting or displaying a monument, though this type of aid is lower in priority compared to the funding of repair work. Instead of grant-aiding work the Secretary of State may empower English Heritage to carry out repair works under s. 5. Powers of entry are also available under s. 6 to ascertain the condition of a monument (which may also be used in connection with reviewing the work associated with a SMC). The owner and/or occupier must be given seven days' notice of the intention to undertake works. However, this power tends to be used only when works are deemed urgently necessary to safeguard a monument as the costs of carrying out such work cannot be recouped from the owner or occupier.

Areas of archaeological importance

The 1979 Act introduced a further protective measure to allow for rescue archaeology to take place before development activity obliterates any archaeological evidence. Under s. 33 an Area of Archaeological Importance (AOAI) may be designated by the Secretary of State (or English Heritage in London) or by a LPA (in its own area) subject to consultation with English Heritage. Since 1982, when the power to designate AOAIs was brought into effect, only five such areas have been confirmed. They cover the historic centres of five cities: namely, Canterbury, Chester, Exeter, Hereford and York. In fact the intention of the provisions was to allow designations to take place in areas of considerable importance rather than in agricultural areas where archaeological remains are more scattered across the country.

Controls and offences in designated areas

Under s. 35 of the 1979 Act it is an offence to cause or permit to be carried out any operations on designated AOAI land involving the disturbance of the ground, or which involve flooding or tipping, without first serving an *Operations Notice* on the relevant LPA and then allowing *six weeks* to elapse before commencing operations. If the LPA is itself the developer, it must serve the notice on the Secretary of State. An offending person may be fined up to the statutory maximum level (see above) on summary conviction or an unlimited amount on conviction on indictment. Defences against prosecution are provided under s. 37(5) and are similar to the defences provided in relation to the unauthorised works carried out to scheduled monuments in that a developer may argue that all reasonable steps have been taken to avoid or prevent the disturbance of the ground or that works were required in the interests of health and safety.

Where an *Operations Notice* is served it should specify the nature of the proposed works and the estimated commencement date and should also certify that anyone with an interest in the land has consented to the works. The Secretary of State will then appoint relevant persons whom he considers competent to undertake archaeological operations (often from the archaeological department of a university or from a local authority) to act as the *Investigating Authority* which then has the right to enter and inspect the site, observe any operations and/or carry out excavations. However, a site may only be excavated if within *four weeks* of the service of the *Operations Notice* the relevant authority serves its own notice on a developer indicating its intention to excavate and serves a copy of the notice on the LPA which has been served an *Operations Notice*. The *Investigating Authority* then has a period of *four months and two weeks,* starting from the end of the developer's *six-week* period of notice, before carrying out any excavations. Thus development may be delayed for a six-month period in total. After six months there is no further statutory bar on any proposed development, which is the subject of a valid planning permission, from taking place.

The relevant form of notices referred to above are detailed in the Areas of Archaeological Importance (Forms of Notice, etc.) Regulations 1984. Section 37 of the 1979 Act allows for certain exemptions to be made from the requirements of s. 35 which are detailed in the Areas of Archaeological Importance (Notification of Operations) (Exemption) Regulations 1984. These include certain operations in relation to agriculture, forestry, mining, tunnelling, landscaping and the repair and maintenance of highways, waterways and mains services, and their installation. Further powers in relation to AOAIs are provided to investigate sites which may be compulsorily acquired (s. 39) and for the *Investigating Authority* to enter sites for the purpose of inspecting, observing, examining, recording during the *six-week* period after an *Operations Notice* has been served and to excavate and take custody of articles following the required *four weeks'* notice being given (s. 38). LPAs also have a power to institute proceedings in the High Court for an injuction to prevent an offence under s. 35 where it appears that an offence is about to be or is being committed.

The future of areas of archaeological importance

The legislative powers relating to AOAI were formulated principally due to a problem that previously existed, which was that the only way to control disturbance of areas thought to be of archaeological interest, apart from scheduling sites, was to attach conditions to planning permissions to enable access to a site during the course of works in order to carry out archaeological investigations. Although the *Hoveringham* case clarified that the archaeological heritage (whether scheduled or not) was a material consideration in determining planning applications, the procedure of using conditions was criticised for two reasons. First, on occasions, important achaeological discoveries have been made when they could not have been reasonably anticipated before excavations had commenced. Thus planning permission may have been granted without conditions because they were not deemed necessary, resulting in the potential loss of any archaeological remains subsequently found. Furthermore, any attempt to schedule remains found in the course of development activity permitted by a permission without relevant conditions attached would raise the question of compensation being paid, as was evidenced by the *Rose Theatre* case. Secondly, the condition-based approach has not always allowed sufficent notification for appropriate archaeological organisations to make their own arrangements to carry out excavation work in reasonable time with the additional consequence of potentially causing undue delay in the commencement of development activity (Suddards, 1988).

However, while the AOAI procedure provides an apparent advantage over the condition-based procedure in allowing an initial evaluation of a site at the application stage, which may reveal a site to be sufficiently important for a planning application to be refused or a discovery to be scheduled, only five AOAIs have been designated since 1982. Moreover, two city councils have been able to reject proposals for AOAIs and a long debate concerning the addition of Oxford to the list of AOAI cities did not result in a designation. Furthermore, despite the fact that the 1979 Act states that a relevant LPA may designate AOAIs, no statutory instrument has been produced to set out the procedures for this (Scrase, 1991).

Yet in 1989, after substantial development activity in central London revealed a number of important discoveries, including the remains of the Shakespearian Rose and Globe theatres and a Roman Amphitheatre at the Guildhall (among others), some consideration was given to the idea of further AOAI designations. Moreover, the Museum of London archaeological service argued that developers would take the isuue of the archaeological heritage more seriously if they were forced to. It was suggested that up to 500 more areas should be designated so that all redevelopment in them should be preceded by four and a half months excavation (Barrie, 1989). This was despite the fact that voluntary contributions by developers towards archaeological investigation had amounted to £14 million in 1988 alone (Sykes, 1991). This had been largely due to the establishment of a permanent body initiated jointly by the British Property Federation and the Standing Conference of Archaeological Unit Managers known as *The British Archaeologists' and Developers' Liaison Group* (BAADLG) which devised a voluntary national code of practice in 1984 (revised in 1988). The preamble to the code of practice had recognised that voluntary cooperation could achieve better results than AOAI designation (Schofield, 1990). However, despite the apparent willingness of developers to cooperate, the increasing number of important discoveries made in the late 1980s and the limitations of protective measures highlighted by the *Rose Theatre* case led to considerable debate on the effectiveness of existing measures.

The outcome was the issuance of a consultative document entitled *Archaeology and Planning* in February 1990 which was followed by the formal release of planning guidance in the form of PPG 16 in November 1990. Paragraph 20 of Annex 3 to PPG 16 indicated that the Secretary of State had decided that no more AOAIs should be designated until the effectiveness of PPG 16 had been assessed. While some commentators have since argued for the protection of ancient monuments to be fully integrated into a planning system, as is the case with listed buildings and conservation areas (the opportunity for this being created by the Planning and Compensation Bill which was enacted in 1991), the provisions of the 1991 Act, which created a presumption in favour of the development plan for determining planning applications (under s. 54A of the principal Act), has arguably dispensed with the need for this (Scrase, 1991).

Protection under planning legislation

Paragraphs 14–16 of PPG 16 indicate that the key to the future preservation of the archaeological heritage is by the development of appropriate planning policies in development plans and their implementation through the development control system. In order to reconcile the need for development with the interests of the conservation English Heritage's advice note to LPAs, *Development Plan Policies for Archaeology*, explains how a framework for the consideration of development proposals that may affect archaeological remains may be written into development plans (English Heritage, 1992). This will be aided by recorded information derived from county SMRs and further information provided by the on-going MPP. The main aim of the new planning guidance on archaeology and development is to bring both developers and LPAs to the attention of matters of archaeological importance at the earliest possible stage so as to avoid potential delay, additional costs, and abortive design work while also being able to determine an appropriate method of preserving ancient remains. This now requires considerable consultation and cooperation before a planning application is actually submitted. Thus, by appropriate planning policies it may be possible to avoid the problems raised by the *Rose Theatre* case.

Development plan policies

In light of the introduction of the presumption in favour of development proposals which are in accord with the development plan in 1991, the DoE has built on the specific guidance produced in PPG 16 by introducing a revised PPG 1 (March 1992) on *General Policy and Principles* to explain the new legislation and a new PPG 12 (February 1992) on *Development Plans and Regional Planning Guidance.* Paragraph 6.5 of PPG 12 gives a clear indication that the archaeological heritage should be given a high priority in development plans. This means that at the strategic level regional planning guidance should take archaeology into account when considering major infrastructure projects such as new motorways and structure plans (or Part I of unitary development plans), should set out a broadly framed policy in favour of preserving nationally important sites and their settings (whether scheduled or not) and indicate a LPA's commitment to promote the appropriate management, knowledge and enhancement of ancient monuments and archaeological sites (English Heritage, 1992).

For example, Berkshire County Council, noted for its leading role in developing appropriate development plan policies in the context of archaeology and planning (Chadwick, 1989), published a new structure plan in 1992, to revise its 1989 approved plan, which gave archaeological interests a high profile. It was felt necessary to make a commitment to archaeological matters due to the destruction of many unprotected sites by new commercial development in favoured locations since the 1960s. Moreover, by 1978 it was estimated that 93 per cent of archaeological and historical deposits in Reading had been destroyed by development activity. The new policy for 1990–2006 redirects retail and housing development to less sensitive areas and otherwise seeks to preserve and manage intact the most significant remaining archaeological sites and to secure the proper investigation of sites of lesser significance, as identified by a maintained SMR and by directing the development of appropriate local plan policies, prior to planning applications being determined (Mortimer and Chadwick, 1992).

English Heritage's advice on the preparation of local plan policies (and for Part II of unitary development plans) has indicated the need to reflect the guidance of para. 27 of PPG 16: namely, that there should be a presumption in favour of preservation of nationally important sites (whether scheduled or not) *in situ* while at the same time the case for preservation should be assessed on the individual merits of each case. In other words the presumption should be against proposals which might cause significant damage or alteration to known remains or to their settings. However, this does not mean that only known archaeological sites should be defined for protection purposes on the proposals map. It has been suggested that plan policies should further take account of previously unknown or undervalued sites with appropriate procedures for dealing with planning proposals at the earliest stage. This is in line with the guidance contained in PPG 16 on the handling of planning applications.

For example, in developing the unitary development plan policies for the five metropolitan authorities within the county of Tyne and Wear four categories of protection policies have been considered in line with the county archaeologist's advice. These include: scheduled sites; sites and buildings which are known to exist and are detectable; areas potentially important due to earlier recorded discoveries or documented evidence; and areas where nothing is known but await field investigation. Moreover, apart from the proposals map identifying specific archaeological sites, an archaeological assessment will be required for large greenfield sites which are substantially 'undeveloped and undisturbed'. Thus, by identifying areas of known or potential archaeological importance, a basis for safeguarding the archaeological heritage is to be established which, subsequently, will be considered in the process of determining planning applications. Further guidance is being provided in the form of non-statutory practice and guidance notes to assist developers and archaeologists in the consideration of matters that may be relevant to the decision-making process (Harbottle, 1993)

Development control policies

Part B of PPG 16 provides a set of principles for giving appropriate consideration to archaeological matters both before and after the determination of planning applications. While these are likely to be justified in development plan policies it must be reiterated that the *Hovingham* case has established that matters of archaeological interest may be a material consideration in the decision-making process.

Paragraph 19 of PPG 16 indicates that the first step in the consideration of development proposals is to have *early consultations between developers and planning authorities.* Pre-application discussions with relevant officals should clarify whether a site is known, or is likely to contain archaeological remains. This may be indicated by the relevant development plan policies. However, where such policies have not yet been formulated, developers are advised to investigate the SMR. Notwithstanding this, consultations with relevant officials concerned with matters of archaeological interest may be needed. The SMR is generally updated through on-going field work by the County Archaeological Officer (and English Heritage in London), and the RCHME is maintaining the information as it is gradually transferred to computerised records in the National Achaeological Record. English Heritage may also provide expert advice outside London. Furthermore, an increasing number of local authorities are now employing their own archaeologists who may be able to offer relevant advice. The object of this pre-submission activity is to alert developers to the potential archaeological sensitivity of a site.

If the outcome of early consultations is that a site is considered to be archaeologically sensitive, para. 20 of PPG 16 indicates that a developer may wish to commission an independent *archaeological assessment.* The LPA or County Archaeologist may be able to provide a list of archaeological consultants for this purpose. An archaeological assessment should attempt to identify the impact of a planning proposal upon the historic environment. It is normally conducted by means of a desktop study of existing information with a field visit of an area to be affected, but will not require a field examination. Sources of information to be investigated may include early maps held by the County Records Office, public libraries, or other archives; aerial photographs held by the RCHME, the LPA, a university archaeology department or other sources; the SMR; and principal secondary sources such as material published in archaeology journals. The assessment should: identify historical features associated with the site, including the presence of any historic buildings, scheduled monuments, other matters of historic interest such as other ancient remains, boundaries, woods, routes, mines or other old workings; describe the potential impact of the proposal upon these items in terms damage, loss or affect on the setting; provide a consideration of other potential matters of archaeological interest based on geophysical survey records, the topography of the site or known patterns of settlement on the site; and, where appropriate, provide means to mitigate the effects of a particular development proposal.

The latter consideration is important as English Heritage has advised LPAs that development plans should include a policy indicating the authority's intention to ensure mitigation by suitable design, including the need for modifications to ensure greater physical preservation where necessary, or by not allowing development to commence until an archaeological investigation and recording exercise has been agreed and carried out (English Heritage, 1992). Moreover, further pre-application discussions following an archaeological assessment which has revealed that there may be important archaeological remains on a site may lead to the conclusion that a *field evaluation* should be carried out. Paragraph 21 of PPG 16 indicates that it is reasonable for a LPA to request a prospective developer to arrange for such an evaluation, preferably by a member of the Institute of Field Archaeologists, details of which are published in Annex 1 to PPG 16.

A field evaluation will normally be conducted by utilising a range of appropriate methods. These may include investigation by trial trenching, geophysical survey, boreholes, aerial photography, field walking, measured surveys, and suitable photographic survey methods for buildings such as rectified photography or photogrammetry. The evaluation should comprise only essential elements. The first element is a description of the surviving archeological remains both above and below ground. This should be illustrated by appropriate methods including the use of maps, plan, elevations, sections and photographs, and cross-referenced to the SMR where possible. The second stage of the evaluation will involve the design of specific measures to mitigate any damage which the proposal may cause to archaeological remains such as altering the proposed foundation design. Apart from consideration of the evaluation in the planning process, a copy of the report should be deposited with the SMR as it may contain new information on archaeological remains.

The results of a field evaluation and any earlier assessment may be required by a LPA before determining a planning application. A voluntary approach to providing information is the underlying basis of the guidance of PPG 16 which builds on the earlier code of practice formulated by the BAADLG code of practice. However, para. 22 of PPG 16 advises that developers can be directly requested to supply relevant information via Regulation 4 of the Town and Country Planning (Applications) Regulations 1988. Moreover, under the Town and Country (Assessment of Environmental Effects) Regulations 1988 the results may be requested as part of a formal environmental assessment. Under Schedule 3, para. 2 of the latter regulations a developer may be requested to submit an *environmental statement* which describes the likely environmental effects, both directly and indirectly, of a proposed development on matters of cultural heritage (among other matters). Whether or not an environmental assessment is carried out, if the results of the archaeological evaluation reveal evidence of important archaeological remains this will be a relevant consideration in the LPA's decision-making process.

The fore-mentioned pre-application preliminary activities have been designed to resolve the problems which have arisen in such cases as the Rose Theatre and others where important discoveries have been made following the grant of planning permission; however, in some instances planning applications may be submitted without prior discussion regarding possible archaeological implications. In such instances, para. 23 of PPG advises that LPAs should consult the SMR and the county archaeologist, who may undertake further consultations with relevant bodies concerned with archaeological matters, and English Heritage. The outcome of these consultations may be that sufficient information is provided to allow a LPA to make a decision. Thus if matters of archaeological interest are revealed from the consultations this is a relevant material consideration which may result in an application being refused. Clearly developers may benefit from taking a cooperative stance at an early stage in the development control process.

The decision-making process can allow for the grant of unconditional planning permission, or with conditions, or refuse permission. Each of these options may be considered at various stages: namely, following a determination that no significant archaeological implications apply to the site, following an archaeological assessment, or following a field evaluation. However, the more that archaeological matters are needed to be assessed, the greater the likelihood that conditional approval or refusal of planning permission will occur.

In some instances planning approval may be obtained by redesigning a proposal in the light of impact assessment studies. In other situations the desired objective of securing the preservation of important remains *in situ* may not be possible. It may still be open to the LPA to grant planning permission, balancing the conflicting public interest issues of the harm to cultural heritage with the need to provide new development, even though the development will destroy important archaeological evidence. In such circumstances a LPA may need to make suitable arrangements for *preservation by record*, i.e. it must be satisfied that the developer will make satisfactory provision for the excavation and recording before planning permission is granted and that this will take place before construction work commences. This may be by means of a voluntary planning agreement under s. 106 of the principal planning Act (as amended), which is likely to involve the use of an agreed development brief prepared by the LPA after taking advice by archaeological consultants. Model agreements on the basis of the BAADLG code of practice may be appropriate in this respect. Although voluntary means of proceeding are often of mutual benefit to the different parties, in some instances the only practicable way of going forward may be by the grant of planning permission subject to conditions.

Planning conditions may be used for different purposes in the context of archaeological preservation. Their use must be in accordance with advice contained in DoE Circular 1/85: The Use of Conditions in Planning Permissions. In other words, they must be fair, reasonable and practicable. The circular suggested a model clause to be used in the case of archaeological sites not protected by the 1979 Act that the *'developer shall afford access at all reasonable times to any archaeologist nominated by a local planning authority, and shall allow him to observe the excavations and record items of interest and finds*'. However, it further stated that development work should not be held up for archaeological investigation. Thus, although designed to allow for a *watching brief* where the archaeological sensitivity of a site has not been clarified, the model clause failed to recognise that recording work may need to be carried out or that further investigation might be appropriate (Hammerson, 1994). However, para. 29 of PPG 16 clarifies that conditions may now allow for a *watching brief* by a nominated archaeologist or specifically to allow for investigation, including excavation work, and recording in the course of permitted operations on site. This approach is advocated by the BAADLG code of practice to enable a *preservation by record* to be undertaken.

A developer may have to consider paying for the costs of archaeological investigation and be aware of what impact this may have on the viability of a scheme. Although English Heritage is able to provide grant aid to fund relevant investigations, such funds are limited. Moreover, preference may be given to non-profit making community-based organisations such as a housing association.

Further consideration will be required concerning the outcome of investigations. Where a *preservation by record* is to be made, responsibility for the production of a published record by archaeological consultants of the archaeological deposits which are unavoidably threatened by development activity will usually lie with the developer. Further cost may be entailed in provision of a suitable recorded archive of information, including the cost of packaging material found in the course of excavations, which may be deposited at a suitable repository such as a museum (Harbottle, 1993).

Consequences for developers

The fore-mentioned procedures have assisted in providing a more efficent means of protecting the archaeological heritage through the planning system. Nevertheless, as development proposals are under greater scrutiny through the development plan and development control systems, with the likelihood of developers having to undertake considerable pre-application activities before planning permission is given, certain precautions may need to be taken. The need for assessment, evaluation, excavation and recording all have an impact in terms of delay, costs and building contractual matters, particularly as there is no guarantee that a site will not be scheduled if an important discovery is made before planning permission is granted. However, a developer who is aware of potential problems can make appropriate provision for them.

Time is the most important cost factor which a developer has to bear in mind when undertaking a proposal involving a site which includes matters of archaeological interest. Delay in completing a project may result in more interest being paid on finance borrowed and inflationary rises on materials and labour. Delay may also have a bearing on the success or failure of a project, as very often a developer will depend on particular market conditions which present an opportunity to supply facilities that are in demand. But markets can change for a number of reasons and a delay may mean that the supply gap is closed by someone else. At the same time many developers and archaeologists are aware that voluntary cooperation can minimise delays (McDougall, 1994). Moreover, the BAADLG code of practice has assisted in creating a more fruitful climate for resolving the needs of each side.

A prudent developer will nevertheless be one who is aware of specific cost implications that archaeology may have on a particular scheme. A number of cost implications should be considered in this respect (Lawrence, 1989).

First, there is the cost of the actual archaeological investigation, including any post-excavation investigations, publication of research findings, archive and display costs. The developer's contribution to these costs should be established at an early stage by negotiation. An appropriate financial sum may be estimated according to comparison of other schemes or the archaeological consultants may be able to estimate the likely costs, including a contingency element. This cost factor will also depend upon whether a watching brief is to be provided or whether extensive access is to be provided for a decided time period. This will largely depend upon the likely importance of the site.

There will also be costs associated with enabling works for on-site investigations. These may include temporary earth work support, temporary support to adjoining buildings, the diversion of mains services, plant and labour for bulk digging, temporary accommodation for site workers, temporary covers to protect working areas from bad weather conditions, and replacement of site material prior to development activity taking place. The main purpose of the enabling works are to allow the investigation to be carried out in a sensible and safe manner. They should therefore be carried out to a high standard and may require an input from a structural engineer, a services engineer and a quantity surveyor.

Apart from the above direct costs other factors may arise which must be considered in the cost equation. As stated previously, the delay caused through archaeological investigation may have a cost implication in terms of interest, inflation and profitability. However, it is important that a developer is aware that archaeology must be considered as part of the development process. The development in planning policy in relation to matters of archaeological interest should alert developers to the consequences of unexpected important discoveries being made during the course of pre-development investigations or after construction work has commenced. However, being confronted with unexpected finds when no provision has been made for this can have a devastating effect, particularly if building contract arrangements have not accounted for this possibility. Yet where it is considered that archaeological remains are likely to be encountered, special provision for pricing consequences of stoppage or delay can be built into bills of quantities or in a separate tender summary on projects based on drawings and specifications. Moreover, the consideration of archaeological matters at the tender stage will force contractors to assess the potential problems and to make their competitive tenders based on their costs. It is likely that this will result in financial benefit to the developer (Davey, 1992). Clearly the conditions and terms of an archaeological agreement should be written into a contract which defines the obligations and rights of all parties involved. This has been considered by the Institute of Field Archaeologists through specific guidance documents (Hammerson, 1994).

The evidence of archaeology may also lead to additional costs arising from the need to alter the design of the scheme to ensure that remains are preserved *in situ,* or incorporated and displayed within a project. The latter approach was used in the Rose Theatre development as a compromise arrangement whereby the remains were protected by redesigning the new building on stilts at a cost of £10 million (Bar-Hillel, 1989). This approach may have a further cost in that in order to house remains for display there may need to be a consequent reduction of income-producing space within a building and with a consequent effect upon the development value of a project.

The cost of archaeology in the development process can be very high. The Rose Theatre case may be contrasted with the discovery of the Roman Baths at Huggin Hill, London, during the course of pre-application investigations which led to the site being scheduled. Despite this, planning permission was granted subject to an archaeological investigation and the remains being preserved *in situ.* The cost of the archaeological investigation amounted to £475,000. However, the resultant delay, and cost of back-filling and sealing the remains, increased the costs by over £3 million (Bar-Hillel, 1989). Nevertheless, it is rare that costs are so high: the location and the importance of the remains resulted in a significant compromise.

In some instances the importance of the finds may result in considerable public funding towards the additional costs. The contribution by the government of £1 million towards the delay concerning the cost of further investigations associated with the Rose Theatre is unlikely to be matched again. Even so, significant contributions may be made where the importance of the remains merit this in public interest terms. For instance, the discovery of the remains of a Norman stone building in Norwich, considered to be of extreme rarity in Norfolk, on a site destined to be occupied by a new Magistrates' Court, was the subject of £65,000 funding largely provided by the National Heritage Memorial Fund, Norfolk County Council, Norwich City Council, Norfolk and Norwich Archaeology Society and the DoE, which allowed for the remains to be preserved in the new building with access for public view (Gregory, 1989).

In a general sense, s. 45 of the 1979 Act, as amended by the National Heritage Act 1983, empowers the funding of rescue archaeology. It enables English Heritage to either undertake, or assist in, or defray the cost of an archaeological investigation in specific projects in relation to land in England which it considers may contain an ancient monument or other matters of archaeological or historic interest. Grant offers are usually made subject to a ceiling, unless other work is agreed to be essential. The publication of a manual entitled *Management of Archaeological Projects* has assisted in setting out good practice guidelines for archaeological project management for recipients of grant aid (English Heritage, 1991). Thus where a development is demonstrably unable financially to make full provision for the level of recovery demanded by the quality of a threatened site, assistance may be offered (Johnson and Taylor, 1991).

Despite the availability of official guidance regarding appropriate steps which should be followed to minimise problems associated with the discovery of archaeological remains, there may still be situations where important remains are unexpectedly found in the course of development projects. Nevertheless, it is open to any developer to consider insuring against such risk. However, insurance cover is not an automatic right. A developer will have to show that all reasonable precautions have been taken to reduce the risk. Thus, at minimum, a desktop investigation would be required as a starting point for analysing the potential risk. Consideration of archaeological matters in building contracts would also be an important factor. If insurance cover is sought it may be limited to cost asssociated with the suspension of work for a limited period, or for the revision of designs where a discovery is so important that it merits scheduling (Hammerson, 1994). In theory it is possible to insure against almost anything, but insuring against archaeological discoveries has been relatively rare until recent times. Yet some insurers have developed special indemnity policies for this risk (Kelvin-Brown, 1989).

Case study: Seven Dials, Bath

The Seven Dials development considered in Chapter 7 also provides a useful case example of a development project on a site with archaeological remains. In fact the site was scheduled as an ancient monument in the late 1960s. Despite this, the development of the now demolished nightclub and garden centre in the 1970s was undertaken without SMC. However, due to the architectural and archaeological sensitivity of the site, in a designated World Heritage city, considerable investigation was required for the new scheme not only in relation to the impact of the proposed development on adjoining listed buildings and the conservation area but also regarding possible damage upon matters of archaeological interest.

The first drawings for the new development were completed in 1988. However, in view of the need to gain acceptance of the proposed design and to gain planning permission and SMC for the development, it was recognised that considerable historical and archaeological research would have to be undertaken. It was significant that the difficulties surrounding the Rose

Theatre case had surfaced before the developer was ready to submit applications for the required consents, alerting them to the importance of preapplication discussions and consultations. This included checking with English Heritage that the site had remained scheduled following the previous nightclub and garden centre development without consent. As a result, planning permission was granted for development on the site in late 1989 subject to a condition that an archaeological investigation would be carried out and that SMC could be granted following the findings of the investigation.

From an initial desktop assessment it was revealed that while Bath was once a walled city, when it acquired its walls was uncertain. It once had a Roman earth rampart, but its ditch had not been found. The possibility of the site containing a ditch of Roman origin, or from a later period, was the reason why the site had been scheduled. The evidence of historical sources first mentioned the ditch in 1189 and it occurred in records sporadically until 1665 when it was 'scoured', after which it occasionally featured in property deeds as a relic feature. Historical sources further revealed the refortification of Bath before 919 AD which possibly involved the rehabilitation of Roman defences. It was also known that an earthen rampart was erected on the line of later medieval walls in the second or early third centuries. Evidence from excavations at a nearby site in 1980 also suggested, but without certainty, that a large ditch had been dug in the front of the city wall or rampart in the fourth century, with the possible existence of a metalled Roman street parallel to the walls and some outworks, possibly of Saxon date (BAT, 1990).

Having accepted the need for an archaeological investigation the developer chose to employ the Bath Archaeological Trust to carry out the work. It was considered that as the city ditch was a linear feature which need only be sampled at relatively few points along its length, a long trench dug archaeologically across the development site would provide an adequate sample of the threatened deposits. The *British Archaeologists' and Developers' Liaison Group* model agreement (shorter version) was used for this purpose, by which the following matters were agreed:

- A nine-week excavation and investigation period, subject to two weeks' notice to confirm the commencement date and requirements regarding the removal of the archaeologists' plant, machinery and other items from the site.
- Requirements as to health and safety and other legal matters associated with occupation of the site.
- Reasonable access for a *watching brief* after the initial investigation but subject to provisions requiring non-interference with development on the site and that 24 hours' notice may be given to complete the work.
- Permission for the archaeologists and their agents to occupy the site by licence only and subject to the prevention of access by any other persons.

- Requirements that artefacts discovered remain in the ownership of the freeholder, except articles discovered by Coroner's Inquest, to be Treasure Trove and that as many as possible would be donated to an appropriate museum or otherwise given or loaned to an approved museum for such periods as necessary for appropriate research and study.
- Requirements that the archaeologists and their agents enter the site at their own risk and indemnify the developer against any resultant claims.
- The payment of a specified donation to the archaeological unit according to a schedule of 'timing and cost' and subject to other conditions including matters relating to premature stoppages of the investigation or a decision not to proceed with the development proposal.
- A requirement that both sides comply with the revised *British Archaeologists' and Developers' Liaison Group* code of practice.

The agreed fee was made according to an estimate provided by the archaeologists for the investigation and included the cost of enabling works with respect to necessary shoring up works in connection with the excavations and the provision of mechanical excavating machinery. The contract allowed for further funding, if the archaeologists could prove the need for it, although in the event this was not necessary with the excavation being concluded after a period of five weeks.

This excavation revealed the ditch to be some 20 metres away from the city wall line, which was adjacent to the development site (Fig 8.1), and concluded that it was likely to be of late Saxon origin. No sign of a Roman defensive ditch was found unless a smaller ditch, found closer to the city wall than the medieval ditch, was one. However, it was considered more likely to be a drainage ditch for a metalled road similar to one that had already been discovered on the north side of the city, thus representing indirect evidence of a Roman defensive circuit. Also found was a scattering of medieval rubbish pits near and over the ditch. One pit was considered to be a saw-pit, taking the history of Saw Close, the name of the street directly to the east of the site, back to the twelfth century. Nevertheless, the main purpose of the archaeologists had been to prove that the ditch was historically cut on three separate occasions in its history, and to date these occasions. This theory was proved (Fig. 8.2), allowing the site to be back-filled, with no major discoveries preventing the site from being developed (BAT, 1991).

The Bath Archaeological Unit had provided relevant advice to English Heritage so that SMC could be granted for the site. However, a *watching brief* was required during the building works which commenced in 1991. As the site had been filled on separate occasions the building foundation was designed mainly by using deep piles. Consequently little was observed. However, the excavation of a sewer trench into Saw Close revealed a near complete section of the city wall (Fig. 8.3). It was noted that the wall survived to within a very short distance of the pavement. From this investigation it was concluded that the wall was almost certainly of medieval date, c. 1300, and

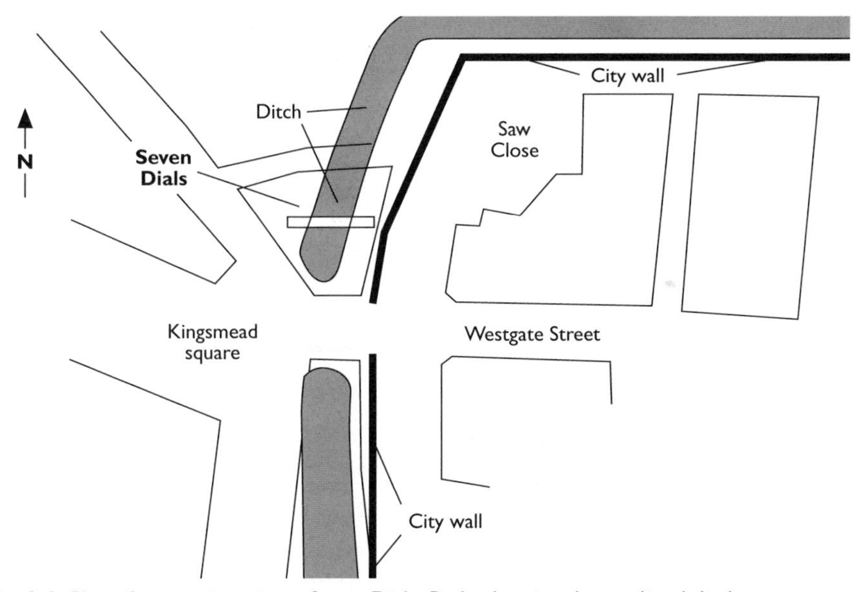

Fig. 8.1 Plan of excavation site at Seven Dials, Bath, showing the medieval ditch. *(Reproduced with the permission of the Bath Archaeological Trust.)*

the rampart probably sealed earlier medieval or Roman remains which would be well preserved at a lower depth (BAT, 1991).

Thus the outcome of the investigations was that there was no need to preserve any remains *in situ,* a historical record having been made of the findings, and that evidence found adjacent to the developent site revealed the need to safeguard archaeological deposits in the surrounding area from any future damage. The development was subsequently completed in May 1992.

Despite initial speculation that the site might have contained important Roman or other significant medieval remains, which might merit the refusal of SMC, the developer decided not to take out indemnity insurance. The view taken was that the risk had been limited by the fact that the proposed development was not for a specific end user. Had it been, and had the developer been contractually bound to provide specific premises, the issue of indemnity insurance would have been more seriously considered. Moreover, the developer had discussed the likelihood of an important discovery with the archaeologists at the desktop assessment stage from which it was concluded that the level of risk did not merit specific insurance cover. Furthermore, certain measures were taken to reduce the risk by undertaking the demolition of the nightclub under a separate contract to the main building contract which was subsequently let after the archaeological investigation had been concluded.

The case study reveals the approach of a responsible developer who was aware of the potential consequences of making a significant discovery on a

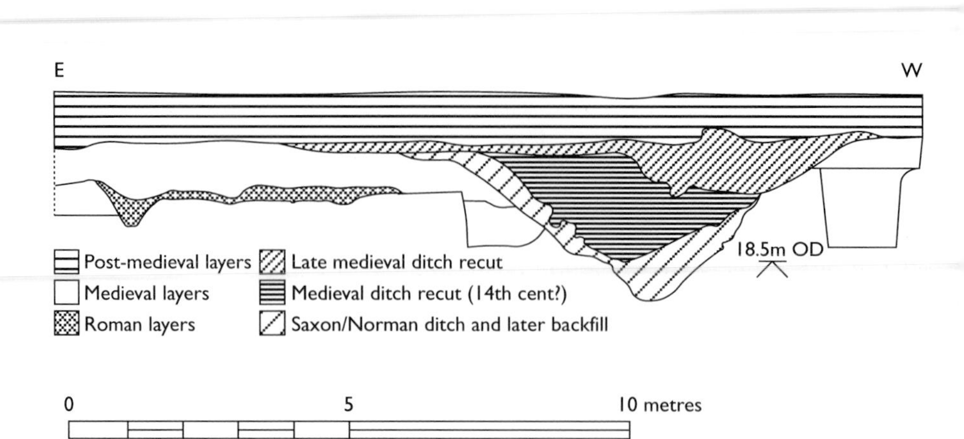

Fig. 8.2 Section of excavations at Seven Dials, Bath, showing layer periods. (*Reproduced with the permission of the Bath Archaeological Trust.*)

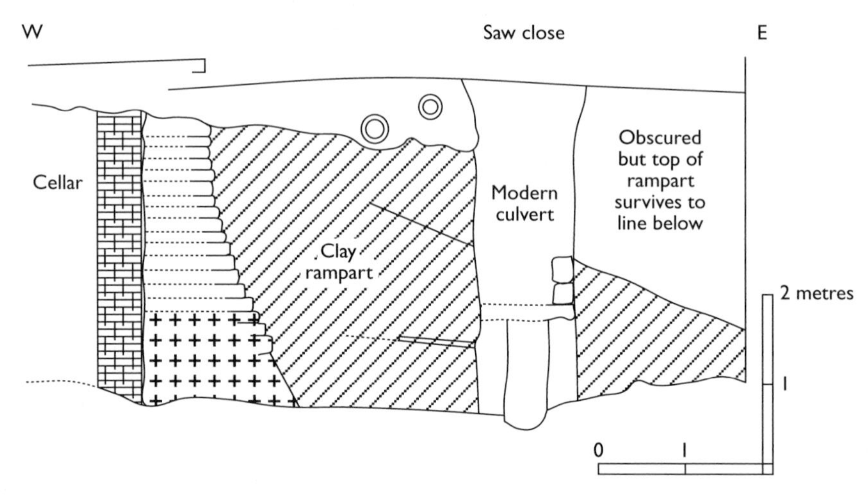

Fig. 8.3 Section through the medieval City defences at Seven Dials, Bath. (*Reproduced with the permission of the Bath Archaeological Trust.*)

scheduled site purchased at considerable expense. A significant amount of risk was entailed in deciding to pursue a development proposal, but this risk was minimised by undertaking early consultations and by working with archaeologists in a cooperative manner, using the voluntary code of practice to guide proceedings. Out of the total development cost of £2.735 million (Pettifer, 1992) the direct costs of archaeological investigation were less than £5000. This successful venture in balancing archaeological and development interests is perhaps even more surprising bearing in mind that the investigations took place before PPG 16 was published in its final form. Nevertheless, it serves as a good case example of the approach now required to be considered by developers working in areas of archaeological interest.

Case study sources

Future Heritage plc (formerly Chartwell Heritage plc and Rosehaugh Heritage plc) (developer)
Arun Evans (project architects)
Tektus (project architects)
Bath Archaeological Trust
Bath City Council (Department of Environmental Services)

References

Barrie, G. (1989) Archaeology: Rose shows failings of code of practice, The Estates Times, 28 July, p. 15.

Bar-Hillel, M. (1989): A Rose by any other name, *Chartered Surveyor Weekly*, 15 June, p. 26.

BAT (1990) Bath Archaeological Trust Annual Report 1990

BAT (1991) Bath Archaeological Trust Annual Report 1991

Brainsby, M. (1994): English Heritage proposed amendments to legislation, *Conservation Bulletin*, Issue 22, March, p. 21.

Carter, H. (1993) Ancient monuments on trial, *Conservation Bulletin*, Issue 20, July, pp. 15–16.

Chadwick, P. (1989) Local government policy – what should it be? Paper presented at a conference entitled *Archaeological Remains and Development*, organised by Henry Stewart Conference Studies, Park Lane Hotel, London, 27 November.

Davey, K. (1992) *Building Conservation Contracts and Grant Aid: A Practical Guide*, E. & F.N. Spon.

English Heritage (1991) *Management of Archaeological Projects*, 2nd edn.

English Heritage (1992) *Development Plan Policies for Archaeology*.

Fairclough, G. (1989) Ancient monuments prosecutions 1989, *Conservation Bulletin*, Issue 9, October, p. 10.

Gregory, T. (1989) Archaeology : so we've found another Roman place, *The Estates Gazette*, 3 June, pp. 20–2 and 62–4.

Hammerson, M. (1994) Building Conservation Note No. 3: Development and Archaeology. An Introduction, prepared with assistance of the RICS Building Conservation Skills Panel and published as a supplement to *Building Conservation Newsletter*, No. 10, Summer.

Harbottle, R. B. (1993) *Practice Note: Archaeology and Guidance Note for Developers and Archaeological Consultants*, Tyne and Wear County Archaeological Office (unpublished).

IAM (1984) *England's Archeological Resource*, Inspectorate of Ancient Monuments.

Kelvin-Brown, J. (1989) The insurance market today. Paper presented at a conference entitled *Archaeological Remains and Development*, organised by Henry Stewart Conference Studies, Park Lane Hotel, London, 27 November.

Johnston, B. (1989) Archaeology guidance promised later in year, *Planning*, 9 June, p. 7.

Johnson, S. and Taylor, R. (1991) Rescue archaeology funding: a policy statement, *Conservation Bulletin*, Issue 14, June, pp. 7–9.

Lawrence, E. (1989): The cost of archaeology on development proposals. Paper presented at a conference entitled *Archaeological Remains and Development*, organised by Henry Stewart Conference Studies, Park Lane Hotel, London, 27 November.

McDougall, L. (1994) Pre-development – digging up trouble?, *The Estates Gazette*, 16 July, pp. 123–5.
Moore, V. (ed.) (1991) Current topics: Scheduled monument consent [1991) JPL 301.
Mortimer, E. and Chadwick, P. (1992) Archaeological sites are given protection, *The Estates Times*, 7 February, p. 23.
Pettifer, M. (1992) Sense and sensitivity, *New Builder*, No. 137, 9 July, p. 21.
Redman, M. (1990) Archaeology and development [1990] JPL 87.
Ross, M. (1991) *Planning and the Heritage – Policy and Procedures*, E. & F.N. Spon, pp. 10–27 and pp. 138–52.
Scrase, T. (1991) Archaeology and planning – a case for full integration [1991] JPL 1103.
Schofield, J. (1990) Digging along with the developers, *Chartered Surveyor Weekly*, 29 March, pp. 138–9.
Startin, B. (1988) The monuments protection programme, *Conservation Bulletin*, Issue 6, October, pp. 1–2.
Suddards, R.W. (1988) *Listed Buildings*, 2nd edn, Sweet & Maxwell, pp. 180–207.
Sykes, D. (1991) Protecting the nation's treasures, *Chartered Surveyor Weekly*, 18 April, p. 94.
Wainwright, G. (1992) Archaeology and planning, *Conservation Bulletin*, Issue 17, June, pp. 23–5.
Wainwright, G. (1994) Archaeology: define first, dig later, *Conservation Bulletin*, Issue 23, July, pp. 10–11.

Chapter 9

SUSTAINING THE HERITAGE

Introduction

It is with hindsight that the mistakes of the *laissez-faire* approach to the planning system which endured through the 1980s have been realised. The policy presumption in favour of development unless it would cause damage to interests of acknowledged importance was not satisfactory in that it could not always be anticipated what effects new development would have. The viability and vitality of many towns and city centres have been undermined by the spate of new out-of-town and edge-of-town business parks, retail parks and centres and superstores. Many historic market towns with a long tradition of independent shops, often passed down through generations of families, have suffered and businesses have had to close with the consequent loss of income filtering through to all aspects of the local economy.

Further debate has scrutinised the way in which development decisions have been made and the consequences of such decisions for the fabric of our towns and cities in the next millennium. A report by English and Overseas Properties on *The Future of British Cities*, published in 1994, was highly critical of the development of corporate office blocks as being inefficient in that they are not fully utilised at night, during weekends and holidays, particularly as improvements in information technology is reducing the need for centralised work places. It further urged the need for investment in the traditional high streets, with a mixture of shops, restaurants, homes and entertainment which may result in a more balanced community. However, the reason why the first direction for institutional investment in property has tended to be away from the historic built environment may be due to the extent of restrictions placed on it.

In recent years there has been a considerable amount of debate concerning the value of the built heritage in both social and economic terms. Some commentators have argued that we have listed too many buildings and taken the protection measures far beyond what was first conceived in the legislation of the 1960s. With nearly 500,000 listed buildings, approximately 8000 conservation areas and about 15,000 scheduled ancient monuments, the extent of safeguards is now considerable, particularly compared to our

European neighbours. Moreover, the scale of protection appears to be ever-widening through the *Monuments Protection Programme*, the extension of the listing process to the more recent heritage, and by a constant growth in designation of new conservation areas. The large number of listed buildings at risk may be as much to do with owner frustration at the inability to be able to alter a building as desired as with the question of redundancy of use. Yet the drive towards sustainable development and maintaining the quality of our urban environments has persuaded the policy makers to a new way of thinking. This seeks to find a new balance between the desire to preserve and the need to create new development in the public interest.

In 1990 the government issued a white paper entitled *This Common Inheritance* (Cm. 1200, September 1990) which introduced new ideas for formulating policies based on the concept of stewardship of our natural and built resources. In 1992 PPG 12 (Development Plan and Regional Planning Guidance) was issued which, at para. 1.8, heralded a new policy based on the concept of *Sustainable Development*, defined by the *World Commission on Environment and Development* (the *'Brundtland Commission'*) in 1987, which seeks to have a more purposeful balance between environmental considerations and the need for new development:

... *The planning system, and the preparation of development plans in particular, can contribute to the objectives of ensuring that development and growth are sustainable. The sum total of decisions in the planning field, as elsewhere, should not deny future generations the best of today's environment. This should be expressed through the policies adopted in development planning* ...

Since this time various planning policy guidance notes have been revised or new policies devised to take account of this new philosophy, including PPG 15, PPG 2 (Green Belts), PPG 3 (Housing), PPG 4 (Industrial and Commercial Development and Small Firms), PPG 6 (Town Centres and Retail Development), PPG 9 (Nature Conservation), PPG 13 (Transport), and PPG 21 (Tourism). In 1994 the government published *Sustainable Development – The UK Strategy* (Cm. 2426, January 1994) which emphasised the role of land use planning, in particular through the development plan system, in implementing policies for sustainability.

The statement that the *sum total of decisions in the planning field ... should not deny future generations the best of today's environment* is important in terms of the built heritage. The words echo the sentiments of John Ruskin regarding historic buildings that '... *They are not ours. They belong to those who built them and partly to all generations of mankind who follow us*', the tenor of which was subsequently reiterated by William Morris in the SPAB *Manifesto* (see Chapter 5). While the statement refers to environmental issues in their widest sense the UK strategy on sustainability nevertheless embraces the built heritage:

... *The built heritage comprises remains of past human settlements, religion, industry, and land use. It includes ancient monuments, archaeological sites, historical buildings and gardens, industrial remains and other landscape features of historic interest. For the prehistoric and many other historic periods, such remains form the only source of evidence for understanding the UK's past. It is, therefore, important that historically and aesthetically important monuments, buildings, sites and landscapes are maintained and protected for future generations. Failure to do so would result in irreversible loss of the nation's heritage ...*

Sustaining the built heritage, therefore, has a wider emphasis on protection than that associated with listed buildings, scheduled ancient monuments and conservation areas. Sites, landscapes and gardens are also to be considered in the context of the historical development of society.

At the same time if the needs for new homes, industry, business and transport are to be met there has to be a re-evaluation of the way in which development decisions are determined. The new emphasis on the primacy of the development plan under s. 54A of the principal planning Act and the movement towards total coverage of land area within England by detailed district-wide local plans allows an opportunity to make appropriate policies for this. Moreover, the joint publication by the Countryside Commission, English Heritage, and English Nature, entitled *Conservation Issues in Strategic Plans,* in 1993 has provided detailed advice regarding how the wider context of environmental concerns should be addressed in the development plan system. This view was also expressed in the 1994 DoE discussion document entitled *Quality in Town and Country* which provided a number of examples of how 'quality' in the built environment can be maintained.

The benefits of a village 'feel' to towns have been highlighted in which opportunities are created for mixed-use developments to revitalise existing communities. A good example to indicate this is in relation to Covent Garden, where 250 buildings were listed in one day in 1973, making Covent Garden one of the highest concentrations of listed buildings in the UK. This prevented plans for the redevelopment of the area into a concrete 'jungle' of high-rise offices, shops and flats linked by pedestrian walkways together with major new roads which had been strongly opposed by the local community. After extensive public consultation a new plan sought to provide more residential accommodation, retain existing buildings where possible, and strengthen the economic character of the area. In other words, to provide a mixture of homes, shops, offices and entertainment facilities to bring new life to the area. Of course this is just history now but it has provided a good exemplar for the future. Specialist retailers and restaurateurs have been attracted and rents for their premises are higher than almost all other locations in the UK. As an office location Covent Garden competes with the City and the West End. It is a lively tourist attraction and most significantly it still has a living community (Deeble-Rogers, 1995).

This approach has also been advocated for town centres. In other words, it is now recognised that more concerted action needs to be taken to maintain the vitality and viability of town centres. This means nurturing the town centre economy by scrutinising the impact of out-of-town developments. For instance, it is now recognised that the development of the Merryhill regional shopping centre in the West Midlands has had a detrimental impact on the vitality of Dudley town centre. It also points to the need for 'town centre management' – a proactive approach which seeks to encourage the right sort of development in the right place, allowing the historic environment to survive while also providing new sensitively sited and designed buildings which complement it. Moreover, it is recognised that the sustainable town will be one which encourages people to live near where they work, which will in turn reduce the impact of the car both visually and in terms of pollution and congestion. Furthermore, traffic management schemes may be implemented in a manner which shows greater sensitivity to the impact of the street on the local environment and counter the demand for the need for new roads that may have a detrimental impact upon the historic environment.

From exemplars elsewhere, much may be learnt about sustaining the built heritage. For example, on the subject of the *Car Free Cities* conference in Amsterdam in 1994, the inhabitants have voted with their feet to remove cars from the historic centre. New corporate business has largely been directed to the outer ring of the city, not affecting the historic core. But the one significant factor here is that people actually live in the old centre. Social housing schemes have been encouraged in the few site opportunities where change can be allowed either in new form in harmony with the surrounding buildings in terms of bulk, height, materials, or by using the shell of historic buildings through flexible re-use policies. The decline in attendance at churches has been even greater than in England yet many historic churches have been the subject of imaginative conversion schemes to turn these buildings into offices, shops, information centres, and housing. Public transport services have been upgraded, but fears that the restriction on car access would affect trade have proved to be unfounded (Palumbo, 1995). Moreover, Amsterdam is a vibrant city with many independent shops, restaurants, cafés, apartments, housing and offices mixed together, and its enhanced canal townscape, historic buildings, museums and galleries provide a strong but managed tourist draw.

The Civic Trust regeneration strategy adopted through the Calderdale Inheritance Project has had a similar beneficial effect in Halifax, and Covent Garden provides another exemplar of the suitability of holistic approaches to sustaining the historic built environment within the wider context of urban renewal. The conservation area partnership approach now being implemented by English Heritage will build on this, but it will require further development of appropriate policies.

There is also a rural challenge in sustaining the social economic fabric of local communities while also safeguarding the natural countryside and buildings in their historic and rural setting. The Countryside Commission guide entitled *Design in the Countryside* published in 1994 has explored ways in which new developments may prevent local character being lost and has encouraged local people to participate in the development control process to ensure that new developments enhance their surroundings and assist in the process of preserving existing sites and settings (Fig. 9.1). The Rural Development Commission also has a major role to play in promoting the economic and social well-being of rural communities by encouraging new enterprise through initiatives which will help to refurbish and re-use redundant farm buildings and industrial sites and also by advising on issues relating to the quality, scale, location and design of new development. English Heritage has commenced a widening of its remit by considering the need for the preservation of historic landscapes.

In general terms it has been recognised that a policy of preservation of the historic environment is not enough. New quality designs for buildings are also needed which both contribute something new to the environment where it is appropriate and also respect existing techniques, use local materials and integrate into the existing context elsewhere. The Secretary of State for the Environment has also reinforced the need for good design which can strengthen community awareness on the basis that 'quality attracts quality' and that 'good design attracts life and investment'. However, he has further concurred that 'it will need considerable vision to meet rising expectations and the demands of the urban economy, yet still safeguard our architectural heritage' (Gummer, 1994).

Sustaining the heritage quality of our environment is a considerable challenge which will require a concerted effort. It is now being pursued by a new policy of sustainability which will take time to be explored through the development of the district-wide plan process in which a more balanced view regarding the new and the old is taken. The protection of the built heritage is now being reassessed in this context to allow greater flexibility to safeguard economic concerns while also bringing additional protection in other areas. Planning policy with regard to the archaeological heritage also requires re-evaluation within the context of 'sustainability' and the on-going review of the ancient monuments. Further consideration is also being given to other issues of the heritage environment, including historic parks and gardens, landscapes, and battlefields, and, of increasing significance, the heritage quality of particular sites and areas is now being reconsidered in an international context through the designation of World Heritage Sites.

Conservation policy approaches

To maintain the built heritage, suitable policies are needed for sustainability, for quality, for saving the best of what we have for future generations. The key to this may partly be in the formulation of an environmentally led development plan system, as has been advocated by the joint statement of the three conservation quangos in their guidance document *Conservation Issues in Strategic Plans*. Through this mechanism it recommends that *key environmental resources*, including heritage resources, should be safeguarded

Fig. 9.1 Fountains Abbey Visitor Centre. A blend of traditional materials and modern design; located in the 'Fountains Abbey, St Mary's Church and Studley Royal Park' World Heritage Site.

for future generations and that demand for new investment in new development should be managed according to the capacity of the built heritage to allow this. However, this will require a survey and analysis of our heritage resource and its capacity to accept change.

In essence this approach would seek to provide a greater understanding of the impact of taking certain actions. Thus, by implication there is a need for more detailed identification of *key heritage resources* and their particular qualities. In other words, heritage matters which are, or have already been, in plan formation may be considered for protection, including ancient monuments, historic buildings, townscapes and landscapes. A review of these features of conservation value should establish how they create character in the environment and their relative merits in social terms as a basis for developing policies in development plans. Thus areas and features which do not fit into the category of a *key environmental resource* are likely to be those where significant and essential change and development can be accommodated.

A parallel process of survey and analysis will also be required to establish the objectives and aspirations of social and economic needs. However, this will need to be on the basis that, wherever possible, new development should not detrimentally affect identified *key heritage resources* or cause a reduction of quality of the wider environmental character. Policy options may require assessment to find the preferred approach by a form of environmental assessment of development concepts. Thus methods of assessment may need to be devised to prevent potentially damaging design ideas for significant individual projects or to encourage the redesign of such projects, and to ensure that policies embrace a balance of development and conservation, or, in other words, sustainable development. In order to do this it may be necessary to utilise a form of assessment of heritage matters in plan formation or in relation to individual buildings to ensure an appropriate balance between protection and flexibility for change.

The need for greater emphasis on character appraisals of areas before the designation of conservation areas is one element of this. Moreover, it is arguable that some areas which may have been designated partly as a response to the desire to attract more funding for heritage matters should be reassessed to see whether the resource actually merits the protection and, if not, whether protection policies should be removed. Furthermore, it has been argued that a clearer method of determining permitted levels of intervention in historic buildings is required and that different approaches should be considered for assessing applications for the public funding of conservation works.

Elsewhere some approaches have been formulated to measure heritage value, the merits of which were scrutinised by Nathaniel Lichfield in his book *The Economics of Urban Conservation* (Lichfield, 1988). One method considered in Canada for the formulation of plans and policies for individual or groups of historic buildings relies on a site survey of buildings assessed according to five main headings. Each of these headings are then subdivided under secondary headings, which are ranked up to a maximum score in each case and overall provide a total of 100 points (Kalman, 1980):

- *Architecture*: including style; construction; age; architect; design; and interior.
- *History*: including person; event; and context.
- *Environment*: including continuity; setting; and landmark.
- *Usability*: including compatibility; adaptability; public; services; and cost.
- *Integrity*: including site; alterations; and condition.

The purpose of this is to find a heritage value ranking against which the opportunity cost ranking (to the owner) can be measured in terms of different options according to the relative points attributed to each issue. These may be idenified as being 'to do nothing', 'to repair', 'to alter for new use', or 'to redevelop'. The highest opportunity cost ranking may well be 'to redevelop', whereas the highest heritage value ranking may be 'to repair'. However, against this should be measured the social opportunity cost. However, this is notoriously difficult to measure and can only really be carried out by assessing certain intangibles on a points system, such as the loss of character of an area to the population at large (Lichfield, 1988).

Another idea considered in the United States has a similar ranking system according to certain defined headings:

- *Historical significance*: including national; state; and community.
- *Architectural significance*: including example of style; importance to neighbourhood; desecration of original design/detrimental additions.
- *Physical condition*: including structure; grounds; neighbourhood; and relation to surroundings.

This form of assessment has been used to rank areas for heritage significance and as a measure against which tax relief, subsidised loans and grants can be given to particular buildings (Catt, 1990).

Another form of assessment, known as the *planning balance sheet analysis*, was used in England as a simulated study in relation to Covent Garden once the alternative of gradual piecemeal renewal of the area and conversion of existing structures within the existing framework of roads and services was accepted. This was measured against the rejected plan for comprehensive redevelopment. The relative costs and benefits of the two schemes were measured in terms of producers/operators (local authorities, central government, developers, business in and out of the area) and consumers (existing residents, future residents, travellers, city population at large and users of new facilities). This showed that while the redevelopment option would have been highly profitable in financial terms there would have been considerable social costs in quantifiable terms of displacement of housing for low-income residents and small business operations, and through congestion costs as well as intangibles such as loss of character of the area (Alexander, 1974). Although the situation has changed considerably since 1974 when this study was carried out, it nevertheless demonstrates the potential for complex areas of heritage value. Moreover, there is no reason why the approach cannot be used for individual buildings to assess the relative merits of different options (Lichfield, 1988).

In England at present only a very basic ranking system is used for listed buildings by grading in the categories of I, II* and II and in relation to conservation areas there is no system of ranking (the category of 'outstanding' conservation area having been removed). But the evidence of the 'buildings-at-risk' campaign of assessing degrees of risk by condition and occupancy has shown up the inadequacies of the grading system, and debate on the designation of conservation areas has arisen because the quality of many designated areas has been eroded. The Conservation Area Partnership initiative has been set up partly in response to these problems. These may encourage more holistic conservation/regeneration schemes on the lines of the experiences provided by Covent Garden and the Civic Trust's Calderdale Inheritance Project. Yet while these may allow for higher funding arrangements, targeting certain problem buildings and improving the environment of some historic areas, it has been recognised that further responses may be required in order to fully address the social and economic balance of issues associated with heritage conservation in the built environment.

In 1994 the DNH, English Heritage and RICS commissioned a research study, undertaken with the University of Reading and DTZ Debenham Thorpe, to examine *The Economic and Social Value of the Conservation of Historic Buildings and Areas* (Page, 1995). The initial results of this research study were announced in June 1995 indicating, from the analysis of different economic pricing methods, that such methods of measuring the relative value of intangible factors may be appropriate for the historic environment. However, further work is needed to analyse successful and unsuccessful case studies to identify the overall contribution of conservation before new policy ideas are formulated (see Appendix 7). Moreover, the Conservation Area Partnership initiatives provide an opportunity to monitor the effects of more holistic approaches to conservation and regeneration over a number of years, which may assist in the development of new policy approaches.

Re-evaluating the architectural heritage

More open discussion concerning the relative social and economic benefits of conservation has led to greater awareness of the different problems associated with conservation policy, as perceived both by owners and investors in property and the conservation lobby. Official recognition of the divide often found between the two sides of the conservation fence is now moving towards a re-examination of built heritage policy.

The announcement in March 1995 by the Secretary of State for National Heritage, that potential cases for listing will be advertised and consultation will be sought from the public and landlords before new listings are made, signified a more open climate in the conservation debate. A swift response from the lobby of property owners and developers was to call for more flexibility in the listing process, including consideration of the possibility of listing only certain parts of buildings such as a façade (Wheatley, 1995; MacDonald, 1995). Undoubtedly this would allow more certainty as to how buildings may be altered in order to ensure their long-term economic use. The RICS also concurred that owners should be able to debate the consequences that listing might have on the economic viability of their buildings (Eade, 1995).

English Heritage has noticeably adopted a more pragmatic approach to building conservation issues in recent years, yet their ability to provide guidance in this area is limited. While para. 1.4 of PPG 15 raised the importance of economic viability in the conservation formula, consideration of methods to give more certainty about the ability to carry out works to listed buildings will undoubtedly require new statutory provisions before they can be implemented. Attempts to create more flexibility at the present time are not possible as LPAs would fetter their discretionary powers by making management agreements to allow alterations to listed buildings without LBC.

However, the Secretary of State for National Heritage also announced his intention to publish a Green Paper covering listings and other heritage matters in late 1995. It was intimated that the main thrust of the listing programme would change towards thematic surveys of buildings offering a better researched basis for identifying, on a selective basis, the best examples of particular building types and periods for listing purposes (Eade, 1995). Whether this will result in changes to legislation allowing greater flexibility for grade II buildings is not certain. Yet there would seem be a movement towards maintaining the strictness of existing controls for listed buildings of outstanding interest (grades I and II*) and also laying greater emphasis on townscape value. This is implied in para. 4.27 of PPG 15, which introduced a presumption in favour of the preservation of unlisted buildings that make a positive contribution to the character or appearance of conservation areas (Carter, 1995). Thus it may be argued that unlisted buildings in conservation areas have gained an increased protection as proposals to demolish such buildings will be judged against the same broad criteria as proposals to demolish listed buildings (paras. 3.16–3.19) (Collinson, 1995). Moreover, the decision to provide greater controls over external alterations to domestic property within conservation areas addresses the importance of façades in the context of townscape value for the public benefit (see Appendices 6 & 7).

Nevertheless, the current debate is to seek avenues for more flexibility and to provide a better balance between the social benefits of conservation and the economic needs of property owners (Page, 1995). Thus re-examination of heritage policy would appear to be moving towards reducing the restrictions placed on alterations to buildings which may have been listed mainly for their external physical contribution to townscape value rather than for their internal features. Greater emphasis on character appraisals for the establishment of new, or the continued designation of existing, conservation areas will ensure a more rigorous view of townscape conservation policy. Furthermore, English Heritage's priorities for funding is to take a more holistic approach to townscape conservation by the replacement of the majority of Town Schemes with Conservation Area Partnerships by April 1997 and in the remaining areas to encourage new partnerships by providing grant aid to formulate action plans.

Undoubtedly the government is seeking to make further changes to the management of heritage policy associated with listed buildings and conservation areas which may result in changes to the current legislation before the commencement of the next millennium. In addition to the possibility of creating more flexibility in the control system associated with listed buildings, English Heritage has also proposed various amendments to the governing legislation concerning grant-making powers, repair powers, the designation of conservation areas and in relation to the meaning of 'demolition' and the difficulties of interpretation as to what forms part of a listed building (Brainsby, 1994).

Re-evaluating the archaeological heritage

The issuing of PPG 16 was undoubtedly partly in response to pressure to clarify what was expected of developers when considering sites with archaeological remains following the *Rose Theatre* case and other important archaeological discoveries in the late 1980s. It accepted the need for development plans to contain policies to safeguard the archaeological heritage which are now to have greater significance within the plan-led planning system.

The *Pagoda Report*, commissioned by English Heritage to research the effects of PPG 16, indicated that its advice had been adopted by every LPA in the country and that by the end of 1991 the archaeological implications of 'virtually all' planning permissions was being properly considered (Pagoda Projects, 1992). However, it is uncertain as to the extent to which this is true as some commentators have questioned the extent of cooperation between district LPAs in allowing County Archaeologists to monitor applications which may have archaeological significance (Brenan, 1994). Moreover, others have argued the case for fully integrating the preservation of the archaeological heritage within the planning system so that the importance of interpreting archaeological sites is not carried out in isolation but rather within the wider historic landscape (Scrase, 1991). At the same time, finding suitable ways of incorporating historic landscape protection within the planning system is not so easy to achieve due to the wide remit of matters which may be inherently worthy of protection (see below).

Furthermore, with the emphasis of PPG 16 being to preserve remains *in situ* the need for excavation is minimised, thus arguably limiting the amount of knowledge which can be gained from investigation of the past. This is compounded by the fact that the onus for paying for excavations has been placed on developers, which inevitably requires a level of compromise on the extent of excavations due to the need to minimise delay and from a cost point of view.

Nevertheless, the review of the archaeological heritage through the *Monuments Protection Programme* may allow the opportunity to develop new approaches to the preservation of the archaeological heritage. For example, to complement this programme, English Heritage, in association with the RCHME, commissioned the Conservation Sciences Department at Bournemouth University to undertake a *Monuments at Risk Survey* (MARS) pilot study. The need for such a survey has been identified from a number of studies carried out over the last decade as archaeological considerations have become more important in terms of environmental assessment, development control and land-use planning and management issues (Wainwright and Darvill, 1994).

The MARS initiative aims to provide a systematic quantification of England's archaeological resource and, in particular, more information on its condition and survival. It will be based on re-examination of the accumulated records in the National Archaeological Record and local Sites and Monuments Records (SMR), and various new research initiatives. The objective of the survey will be to develop a database which will be added to that of SMR and also used for further analyses at national, regional and local levels to provide a general understanding of the dynamics of the archaeological resource.

In association with the findings of other countryside and landscape studies this will enable further initiatives to be developed for the management of the resource and to enable decisions regarding its protection to be better informed. It may also be used to target resources under the MPP towards monuments and historic landscapes most at risk. Clearly much survey work is yet to be carried out before any significant changes of policy can be introduced. Yet the eventual outcome of current initiatives is likely to produce amendments to existing statutory provisions. Moreover, English Heritage has made various proposals for amendments to the Ancient Monuments and Archaeological Areas Act 1979, including new criminal offences and greater penalties for offences, a widening of the definition of a scheduled ancient monument to give greater protection including more consideration of a monument's setting and environs, and the introduction of enforcement powers similar to those applied to listed buildings (Brainsby, 1994). Whether such proposals are adopted and whether there will be greater integration of provisions for the protection of the archaeological heritage within the planning system remains to be seen.

The wider historic environment

Government policy towards the historic environment is in the process of widening to take a more holistic view, including recognition of the contribution made by historic parks and gardens, landscapes and battlefields, as well as the designation of World Heritage Sites (WHS). The importance of these aspects is recognised in PPG 15, which stresses the role of the development plan system in their protection and preservation through the formulation of specific planning policies. The protection of these issues is also a material consideration in the determination of planning applications which may affect them.

World heritage sites

Under Article 5(a) of the UNESCO *World Heritage Convention* 1972 signatories to the convention have made a commitment 'to integrate the protection of the (cultural and natural) heritage into comprehensive planning programmes'. Each signatory state is responsible for nominating WHS which are assessed first by ICOMOS in terms of *outstanding universal value* and then by the World Heritage Committee elected by the signatory states. There are now over 300 heritage or man-made sites which have been registered as WHSs.

The convention was ratified by the United Kingdom government in 1984 and since this date ten WHSs have been designated in England. These are: Fountains Abbey, St Mary's Church and Studley Royal Park (Fig. 9.1); Durham Cathedral and Castle (Fig. 9.2); Ironbridge Gorge; Stonehenge, Avebury and Associated Sites (Fig. 9.3); Blenheim Palace and Park; Palace of Westminster and Westminster Abbey; City of Bath (Fig. 9.4); Hadrian's Wall Military Zone; The Tower of London; and Canterbury Cathedral, St Augustine's Abbey and St Michael's Church. The areas of WHS designation have often been chosen to coincide with conservation areas or scheduled ancient monuments. The nomination of the Lake District National Park has been deferred until the position regarding 'cultural landscapes' has been clarified by the committee (Evans *et al.*, 1994). One further WHS nomination has been recommended by the *Royal Parks Review* to include Greenwich Palace and Park which features works of a number of prominent British Architects including Inigo Jones, Sir Christoper Wren and Nicholas Hawksmoor (Brown, 1995)

Until PPG 15 was issued in September 1994 it was unclear as to the status afforded to WHS in England. Nevertheless, the importance of WHS designations has been reflected in a number of planning appeal decisions:

The inspector at the public inquiry concerning a proposal to carry out works by Marlborough Homes at West Kennet Farm, Avebury, confirmed that the WHS designation drew attention to particular interests of acknowledged importance but did not impose a further layer of control, and WHS status did not mean that there could be a rigid ban on development. Nevertheless, the decision letter of the Secretary of State acknowledged the fact that WHS listing meant that the Avebury complex was 'not only of

Fig. 9.2 'Durham Cathedral and Castle' World Heritage Site.

Fig. 9.3 Stonehenge, England's most famous scheduled ancient monument; located in the 'Stonehenge, Avebury and Associated Sites' World Heritage Site.

Fig. 9.4 The Royal Cresent, Bath. An essential part of the only World Hertage City in England.

national but international importance' (Ref. SW/P5407/21/74 and 270/199 and HSD/9/2/1517Pt. 3 and 2004). However, in a further proposal within the Stonehenge and Avebury WHS concerning an application for a hotel/hostel the inspector only briefly mentioned the WHS and the resultant decision letter did not refer to it all (Ref. SW/P/5407/21/67).

The Hadrian's Wall WHS was subject to two controversial planning applications. In an application which proposed test drilling for hydrocarbons in close proximity to Hadrian's Wall the setting of the WHS was the major issue but the application was actually refused for other reasons (Ref. M42/R2900/01). However, in another application to extract opencast coal only a short distance from the previous site while it was recommended for approval by the inspector, the Secretary of State felt that insufficient weight had been given to the WHS designation. This led to a High Court challenge on the grounds that the Secretary of State had rendered his reasoning obscure by making the WHS designation a major consideration. In *Coal Contractors Ltd v. Secretary of State for the Environment and Another* [1993] EGCS 218, it was confirmed that while circulars or PPGs did not deal with WHS designation, the views by the Secretary of State were. Of note, a Ministerial statement in the House of Lords six weeks before the judgment had confirmed that WHS designation was a material consideration in planning decisions (Baroness Trumpington, 1993).

Following this case a clutch of applications for superstores proposed in the City of Bath were decided by the Secretary of State in 1994 (Ref. SW/P/5116/220/3 and 4 and APP/P0105/A/92/211437). In each case the special importance of Bath, as the only complete city in the country designated as a WHS, was considered. The decision letters which rejected the proposals all remarked that while no new additional planning restictions followed from WHS designation, rather more weight than normal was to be attached to relevant policies of the replacement city plan which had been framed in recognition of the city's 'very special importance as a world heritage site'. Thus, by implication, the relevance of s. 54A of the principal planning Act was reflected in these decisions as well as the factor of the WHS designation being a material consideration.

The need to address the question of WHS designation in the context of planning policy became apparent from these cases and a number of others involving proposals to develop tennis courts in the Durham WHS (Evans *et al.*, 1994), to erect a new bridge at the Ironbridge Gorge WHS (Ref. WMR/P/502/223/6) and to make a *Byway Open to the Public* at Studley Royal Park WHS (Ref. FPS/D2700/7/41). This was provided for in paras 2.22 and 2.23 of PPG 15. This emphasises that while no new statutory controls follow from the designation of a WHS, the designation is nevertheless a *key material consideration*. Moreover, each LPA with a WHS designation was requested to formulate specific planning policies for the protection of these sites and to include them in their development plans, the emphasis being placed on their *sustainability* to protect them for future generations. While development proposals may be compatible with this objective, any proposal affecting a WHS or *its setting* are to be carefully scrutinised, and in the case of significant development proposals it has been advised that a formal environmental assessment will be required to consider whether the immediate impact and implications for the future are fully evaluated.

Furthermore, para. 6.37 encourages LPAs to work with owners and managers of WHSs in their areas, and other relevant agencies, to ensure that comprehensive management plans are formulated for them. The purpose of these plans is to appraise the significance and condition of the site; to ensure the physical conservation of the site to the highest standards; to protect the site and its setting from damaging development; and to provide clear policies for tourism as it may affect a site. However, as WHSs are a new phenomenon for LPAs to deal with, LPAs have been directed to seek the advice and assistance of ICOMOS.

One of the probems highlighted from the planning cases involving WHSs mentioned above was that boundaries of WHSs have tended to follow existing boundaries such as for conservation areas or scheduled ancient monuments but, by example, the 'Military Zone' of Hadrian's Wall implies that further areas may be of interest (Evans *et al.*, 1994). Moreover, initial monitoring reports on all WHSs by ICOMOS UK (with the exception of Hadrian's Wall WHS) have further highlighted a number of common features which need consideration in the development of management plans. These include the definition of boundaries and buffer zones; the formulation of appropriate methods for the systematic inspection, maintenance and repair of features of interest within a WHS; funding mechanisms for repairs; and approaches to dealing with issues relating to tourism and the effects of traffic (ICOMOS UK, 1995). Funding for this appraisal work has been provided by the DNH and private sources and has led to substantial cooperation between ICOMOS UK, relevant LPAs and other agencies to ensure that management plans will be developed in the short term.

In the case of Hadrian's Wall WHS, English Heritage is taking the lead role in coordinating all agencies concerned with the wall zone to prepare a management plan. This is necessary because 13 local and national parks authorities are involved as planning authorities in the WHS which extends from Newcastle to Carlisle (but does not include these urban areas) and ownership of land is spread between English Heritage, the National Trust, local authorities, private trusts and numerous private owners of agricultural land. Moreover, the zone has no closely mapped boundary but includes numerous scheduled and unscheduled ancient monuments and archaeological remains as well as the wall itself and associated Roman forts. It is proposed that the management plan will address a number of important aspects including planning frameworks and statutory controls which can or should apply to the wall zone; a code of conduct for agricultural activities; and visitor management within the wall zone. Further research and investigation of the wall zone in conjunction with the MPP will check and revise scheduling of ancient monuments within the WHS and the RCHME is to provide a database as part of the management plan (Johnson, 1994).

Historic Parks and Gardens

Historic parks and gardens are more easily definable than historic landscapes, hence the decision to establish a *Register of Historic Parks and Gardens of Special Historic Interest in England*. This was first mooted by Lord Montague of Beaulieu, as chairman of English Heritage, in 1984 (Montague, 1984). In fact the National Heritage Act 1983 amended s. 8 of the Historic Buildings and Ancient Monuments Act 1953 to empower English Heritage to compile a register of gardens and land of special historic interest.

However, the initial motivation for compiling a register had grown out of concern over the fate of historic parks and gardens which, since the 1960s, had been increasingly threatened by development pressures from new roads, housing and golf courses in particular. Moreover, the planning system has had a limited effect in preventing the erosion of this aspect of our heritage as many parks and gardens have been affected by numerous small-scale activities which lie outside planning control, such as the construction of small buildings and removal of paths and hedges. Agricultural development and activity has also had an impact on park landscapes. Further damage occurred through lack of proper management by owners (Jacques, 1991).

The Garden History Society (GHS) had a significant role in demonstrating the importance of historic parks and gardens and producing unofficial lists in the 1970s, and other organisations such as the County Gardens Trust (CGT) carried on this work until the decision to establish an official register was made. The *Register*, which was compiled between 1984 and 1988, included only nationally important sites and was published on a county by county basis. The completed *Register* included a total of 1085 sites. However, in 1989 it was decided that a comprehensive review was required, which is set for completion in 1999. By 1994 additions brought the total registrations to approximately 1200, and it is anticipated that this will rise to 1500 registered sites by the end of the review period (Roberts, 1995a).

Over a hundred registered sites are located in London but otherwise there is a spread of sites of varying types throughout the country. English Heritage has classified sites in three broad groupings, including those associated with domestic use, those associated with institutions and those designed for public amenity. However, of more significance, a grading system has been adopted on the same basis as the gradings associated with listed buildings (but entirely independent). Thus when the *Register* was first completed, 10 per cent were classified as grade I (of exceptional historic interest), 30 per cent were classified as grade II* (of great historic interest) and the remaining 60 per cent were classified as grade II (of special historic interest). Also, similar to listed buildings, certain *criteria for selection* have been established:

1. *Parks and gardens formed before 1750, where the original layout is still in evidence.*
2. *Most parks and gardens laid out between 1750 and 1820 if they still reflect the intentions of the original layout.*
3. *The best parks and gardens laid out between 1820 and 1880 which are in good or fair condition and of aesthetic merit.*
4. *The best parks and gardens laid out between 1880 and 30 years ago which are in good condition.*

Where resources permit a more detailed site assessment, particular attention may also be paid to further criteria:

5. *Parks and gardens which were influential in the development of taste, whether through reputation or reference in literature.*
6. *Parks and gardens which are early or representative examples of the genre layout or the work of a designer of national stature.*
7. *Parks and gardens having an association with significant historical events or persons.*
8. *Parks and gardens with group value, especially as an integral part of a layout surrounding a major house or as a part of a town planning scheme.*

(*English Heritage, 1992*).

The pattern of design development reflected in these criteria roughly corresponds to the work of designers of national importance. Some designers, such as Lancelot Brown and Humphrey Repton, were particularly influential, with their names being associated with registered sites on 122 and 103 occasions respectively (Roberts, 1995a).

A further similarity to the regime of listed buildings (and conservation areas) exists in that a site must be of at least *special interest in a national context* to merit registration. However, the similarities do not extend to statutory protection. In fact the *Register* does not imply any additional powers to control development or other work to such sites. Nevertheless, the historic interest of a park or garden has been established as a material planning consideration which is recognised in para. 2.24 of PPG 15, and the *Register* provides the means of identifying sites of special historic importance. Moreover, once a site has been added to the *Register*, the owner, occupier(s), the Secretary of State, and the county and district planning authorities are informed and then sent a copy of its entry. LPAs are also sent a map which defines the boundary of registered sites which may be periodically revised as historical knowledge increases through research. Emergency registering of further sites in advance of the existing review programme may also be considered where research identifies their relative merits and where they are considered to be under threat (English Heritage, 1992).

English Heritage will provide advice, when requested, on any planning or highway proposal which may affect registered sites or their settings. However,

LPAs have had the main role in ensuring the protection of registered sites until recently. In 1995 a DNH press release announced that, following a House of Commons National Heritage Committee recommendation, formal arrangements are to be introduced to require LPAs to consult English Heritage on all development affecting grades I and II* registered sites and to consult the GHS in respect of all registered sites (DoE, 1995). Moreover, they have been advised that some sites not on the *Register* may also have significant historical interest and will in many cases be worthy of conservation.

Apart from the protection of registered sites being a material consideration, LPAs have also been advised to protect registered sites through appropriate development plan policies. In the countryside the protection of historic parks and gardens, often associated with country houses or estates, will be relatively easy to achieve; but in urban areas, where there is pressure to develop a limited number of open sites, detailed planning is necessary to safeguard parks and gardens not just because of their historical importance but also because of the need to retain the amenity value of attractive open spaces (Roberts, 1995b). It is apparent that most revisions of local plan policies in recent years have sought to include policies relating to historic parks and gardens which, in general, state a presumption against development of these sites. However, there will also be a need for policies to widen the scope of protection through development plans by ensuring that the ambience of such sites is safeguarded in the context of their setting such as by avoiding intrusion into views off-site and disturbances from the development of new roads (Jacques, 1991). While section 5 of PPG 15 emphasised the need to mitigate the impact of new traffic routes on the historic environment, para. 5.20 of PPG 13 (Transport) and the Department of Transport's *Highways Manual* specifically mentioned the need to safeguard registered sites.

Yet the safeguarding of registered sites is not just to be managed through planning systems. It has been recognised that active conservation work is also necessary. In this respect English Heritage has the power to provide grants for the repair of gardens and other land of outstanding quality (listed grade I or II* in the *Register*). The DoE has also provided financial assistance for repairs following major storm damage in 1987 and 1990. A pilot scheme to assist more general restoration works to parks and gardens was launched in 1991. Much wider assistance was provided in 1993, coinciding with the funding theme of the European Commission's Cultural Department (DG-X) in the same year to conserve historic parks and gardens under the *Pilot Project for the Conservation of the European Architectural Heritage.*

Under the new scheme grant aid became eligible for the restoration of features that contribute to the *special interest* of registered sites including terraces, steps, water features and tree planting. Assistance has also been provided to prepare restoration and management plans, the intention being to encourage owners, whether private owners, local authorities, or trusts, to make a long-term commitment to site management. Some of the first of these schemes include works for restoration of Charles Hamilton's landscape park at Painshill Park, Surrey; Charles Barry's Italiante partere at Harewood House, North Yorkshire; lime avenues at Ham House near Richmond upon Thames and at Lyme Park near Stockport; and the landscape setting of Prior Park, which lies on the edge of the WHS at Bath, and at Farnborough Hall in Warwickshire. Due to the limited amount of resources available to English Heritage for such schemes, close liaison with potential applicants is required and the advice of the advisory *Historic Landscapes Panel* may be sought before committing funds. Opportunities to dovetail grant aid and other funding sources may also be utilised such as from the European Union or the Countryside Commission's *Countryside Stewardship* scheme (Stretton and Bilikowski, 1993).

More recently the Heritage Lottery Fund criteria for identifying specific categories of eligible assets for support includes repairs to and restoration of parks and gardens, landscape works, and the repair of garden buildings and follies, graveyards, funerary monuments and cemeteries. In the first instance English Heritage is promoting the restoration of public parks, garden squares, churchyards and cemeteries in London through this opportunity. The idea seeks to attain greater awareness of the need for stewardship of these historic sites and to improve public access, enjoyment and understanding of them (Davies and Edgar, 1995). The Heritage Lottery Fund provides an unprecedented opportunity to make an impact on registered sites within London which will undoubtedly be extended elsewhere in time.

Historic battlefields

In 1993 English Heritage announced that another unique register to highlight the country's battlefields would be devised following research carried out by the National Army Museum and the Centre for Environmental Interpretation in Manchester (Lindsay, 1993). The importance of England's battlefields has been highlighted in relation to political history, and in terms of education, tourism, and recreation potential. The fact that many have survived to a remarkable extent is also significant in terms of our cultural heritage.

Battlefield sites are to be chosen for their political, military or biographical significance. The reliability of evidence for each battle will also be a contributory factor, including evidence of documents, archaeology, topography and landscape history. Landscape evolution will also be an important component of battlefield interpretation and presentation in the register (Brown, 1994).

Conservation of battlefields will necessarily involve management. Four important factors have been identified in this respect. First, that large-scale changes to battlefield topography should be avoided to retain authenticity.

Secondly, that visual amenity of battlefields should not be diminished by inappropriate elements in the landscape such as new buildings sited in key views, tree plantations and road embankments. Thirdly, that ground disturbances should generally be avoided due to possible archaeological remains which may exist while accepted methods of archaeological and historic research may enhance the integrity of battlefields. Lastly, that the need for accessibility does not jeopardise the interests of owners and occupiers of sites and, more significantly, that conservation of the educational and amenity value of battlefields is ensured for future generations by the planning system (English Heritage, 1994).

As with the *Register of Parks and Gardens*, the *Register of Historic Battlefields* will not provide any additional statutory powers of protection. However, the implications of the register have been recognised by the government in PPG 15, which indicates at para. 2.25 that the effects of any development on registered battlefields is a material consideration to be taken into account in determining planning applications and, as an identifiable *key heritage resource,* it is likely to be considered in the formulation of development plan policies. From a highways point of view, para. 5.20 of PPG 13, as in the case of registered historic parks and gardens, emphasises that local authorities should avoid the impact of new road schemes on registered battlefields. Moreover, while English Heritage will maintain and review the register, local authorities have been identified as being best placed to deal with the management and conservation of battlefields. This will be via a dual role of regulating development affecting sites and, in association with other agencies, enhancing the potential of battlefields for educational and tourism purposes (Brown, 1994).

The register was formally adopted in 1995 following a consultation exercise which ended in December 1994. At the consultation stage 43 battlefields were identifed for proposed inclusion in the register (Fig. 9.5). Other sites may be added if sufficient evidence is found to merit inclusion in the register following a period of review.

Historic landscapes

In June 1991 English Heritage published a brief policy statement on historic landscapes following the Department of Environment's invitation to prepare another register, namely *a Register of Historic Landscapes,* in the 1990 White Paper *This Common Inheritance.* This highlighted the varied historic components of the landscapes shaped by human use as including those which form part of the land (hedges, walls, woods, fields, tracks) and individual features (earthworks, ruins, barns, and settlements) (Fig 9.6). The premise for this initiative was simply that the physical remains of the past make a contribution to the appearance of the landscape in the same way that they contribute to the appearance of the townscape and are therefore worthy of conservation (Fairclough, 1991).

Fig. 9.5 Monment to the Battle of Flodden Field. One of the sites included in the Register of Historic Battlefields.

Fig. 9.6 Traditional stone barn and dry-stone walls in the historic landscape of the Northern Pennines.

Examples of historic landscapes have been illustrated by English Heritage and the Countryside Commission, for instance, in Swaledale, North Yorkshire, which has the distinctive settlement and field patterns of the Pennine Dales with hay meadows, enclosed upland grazing, and dispersed farmsteads with field barns; near-complete medieval deserted villages and field systems preserved within enclosure hedges in Warwickshire; and in the distinctive English lowland settlement pattern of the Blackdown Hills (Fairclough, 1991, 1994). However, the initial work towards developing a *Register of Historic Landscapes* recognised the need to define what constitutes a *historic landscape* and the precise form and scope that a register may take. At the outset certain broad principles were set out for the establishment of a register. These included the need to cover *all* historical elements of the countryside; the weighting of landscapes in terms of importance, possibly on a national and local basis for prioritising resource allocations; the development of a methodology for defining and grading historic landscapes and the man-made features within them; and a process for informing and assisting management and conservation decisions.

Responses to the policy statement on historic landscapes from a wide variety of individuals and organisations were significantly diverse. English Heritage then designed a programme to look more closely at some of the issues raised. A historic landscape project, devised in association with Kent County Council, demonstrated the need to focus priorities on features rather than areas to measure archaeological and historic value within the countryside. English Heritage also provided an input into a pilot study conducted by the Countryside Commission to examine the division of the country into areas of homogeneous and distinctive character by sponsoring additional work aimed at identifying historic character. Further projects were implemented to experiment with methods of assessing historic landscape importance; to review historic landscape work in farm or estate surveys; to identify objectives and uses of historic landscape assessment for conservation, planning, agriculture and countryside policy; and to define a framework for a formal process of historic landscape assessment (Fairclough, 1994).

An increasing amount of partnership between English Heritage, English Nature and the Countryside Commission has also occurred in relation to the development of policy ideas for the countryside (Fairclough, 1993, 1994) to ensure that planning and sustainable development policy ideas for the countryside are not isolated. This was reflected in the joint guidance document *Conservation Issues in Strategic Plans* and PPG 15 which indicates that the components and character of the wider historic landscape are to be effectively integrated into the planning system. However, para. 6.40 now indicates that the concept of a definitive national register of England's wider historic landscape is now unlikely to come about following the findings of the various research initiatives highlighted above. Furthermore, para. 2.26 advises LPAs, in defining planning policies for the countryside, to take account of the historical dimension of the landscape as a whole. It has been determined that with the whole of the landscape being 'an archaeological and historic artefact' in varying degrees a more holistic policy view is needed to ensure that the balance between conservation and strengthening the rural economy with sensitive development is maintained. With the development of suitable development plan policies, the primacy of the development plan via s. 54A of the principal planning Act may be used to protect the most important components and allow development that assists in maintaining the overall historic character.

While the concept of a *Register of Historic Landscapes* is unlikely to be forthcoming, the Countryside Commission's document *Landscape Assessment Guidance* published in 1993 and continuing work by English Heritage on methodology for historic landscape assessment will assist in the formulation of appropriate development plan policies and management strategies to sustain the countryside heritage.

The national commitment to conservation

As a final note to this chapter it must be fairly stated that the government's commitment to the historic environment is now very significant. This has been emphasised since the publication of *This Common Inheritance* in 1990 and through the document *Sustainable Development – The UK Strategy* published in 1994 and more directly through the revision of PPGs which assist the process of environmentally led planning to be featured in the development plan process. All PPGs are being revised to encompass sustainable development issues and PPG 15 in particular has made a significant contribution towards developing a more holistic approach to the conservation of the elements which make up the historic built environment. Reiterating the definition of the UK Strategy document ...

... *The built heritage comprises remains of past human settlements, religion, industry, and land use. It includes ancient monuments, archaeological sites, historical buildings and gardens, industrial remains and other landscape features of historic interest ...*

Thus policy commitments have been made in terms of funding, practical conservation work, and managing the existing historic built environment encompassing the widest range of aspects. The process of review of the built heritage can now be undertaken on the firm footing that the practical holistic framework has been established to protect it. Further measures may be introduced to bring additional safeguards with the possibility of a statutory mechanism of protection for the *registers* associated with historic parks and gardens and battlefields. Yet the emphasis of conservation thinking is perhaps more likely to move from the fabric to the community in the next millennium.

References

Alexander, I. (1974) City centre redevelopment: an evaluation of alternative approaches, *Progress in Planning*, 3 (1) (edited by D. Diamond and J.B. McLoughlin), Pergamon.
Baroness Trumpington (1993) *Hansard*, 21 October, p. 702.
Bevan, J. (1994) PPG 16 and the restructuring of archaeological practice in Britain, *Planning Practice and Research*, Vol. 9, No. 4, pp. 395–405.
Brainsby, M. (1994) English Heritage proposed amendments to legislation, *Conservation Bulletin*, Issue 22, March, p. 21.
Brenan, J. (1994) PPG 16 and the restructuring of archaeological practice in Britain, *Planning Practice and Research*, Vol. 9, No. 4, pp. 395–405.
Brown, A. (1994) Taking up arms for battlefield conservation, *Conservation Bulletin*, Issue 24, November, pp. 3–4.
Brown, P. (1995) £2m cut blamed for 'shabby' royal parks, *The Guardian*, 1 March, p. 9.
Burman, P., Pickard, R.D., and Taylor, S. (ed.) (1995) Papers presented at a consultation entitled The Economics of Architectural Conservation, held at the Institute of Advanced Architectural Studies, University of York, in association with the European Union Department DG-10, 12–14 February.
Carter, H. (1995) Planning and the historic environment, *Conservation Bulletin*, Issue 25, March, pp. 18–19.
Catt, R. (1990) *Valuation of Historic Buildings*, RICS Diploma in Building Conservation, The College of Estate Management, Reading.
Collinson, I. (1995) Conservation implications, *Property Week*, 9 March, p. 30.
Davies, P. and Edgar, J. (1995) Lottery aid for London's historic landscapes, *Conservation Bulletin*, Issue 25, March, pp. 1–2.
Deeble-Rogers, R. (1995) Covent Garden – an historical perspective, *The Estates Gazette*, Issue 9502, 14 January, pp. 127–9.
DoE (1995) Department of the Environment Press Release No. 94/1995.
Eade, C. (1995) Landlords to get a say on listing of buildings, *Property Week*, 16 March, p. 10.
English Heritage (1992) *The Register of Parks & Gardens* (Advice Note).
English Heritage (1994) *Battlefields: The Proposed Register of Historic Battlefields* (Guidance Note).
English Heritage (1995) *Developing guidelines for the management of listed buildings.*
Evans, D.M., Pugh-Smith, J. and Samuels, J. (1994) World Heritage Sites: Beauty contest or planning constraint [1994] JPL 503.
Fairclough, G. (1991) Historic landscapes, *Conservation Bulletin*, Issue 14, June, pp. 4–5.
Fairclough, G. (1993) Natural partners: liaison with English Nature, *Conservation Bulletin*, Issue 21, November, p. 24.
Fairclough, G. (1994) New landscapes of conservation, *Conservation Bulletin*, Issue 22, March, pp. 16–17.
Gummer, J. (1994) *Quality in Town and Country: A Discussion Document*, Department of the Environment.
ICOMOS UK (1995) *World Heritage*, ICOMOS UK Newsletter, March, p. 3.
Jacques, D. (1991) Planning for parks and gardens, *Conservation Bulletin*, Issue 13, February, pp. 12–13.
Johnson, S. (1994) A management plan for Hadrian's Wall World Heritage Site, *Conservation Bulletin*, Issue 22, March, pp. 4–5.
Kalman, H. (1980) *The Evaluation of Historic Buildings*, Ministry of Environment, Ottawa.
Lichfield, N. (1988) *The Economics of Urban Conservation*, Cambridge University Press.

Lindsay, J, (ed.) (1993) Battle stations for register, *English Heritage Magazine*, Issue 24, December, p. 8.

MacDonald, M. (1995) Secret heritage listings to end, *The Independent*, 9 March, p. 3.

Montague, Lord (1984) The conservation of historic gardens. Proceedings of a Symposium held by the Garden History Society and the Ancient Monuments Society, 9 May, pp. 1–2.

Page, J. (1995) Can we put a value on the heritage?, *Conservation Bulletin*, Issue 25, March, pp. 14–15.

Pagoda Projects (1992) *Pagoda Report: An Evaluation of the Impact of PPG 16 on Archaeology and Planning* (compiled for English Heritage).

Palumbo, Lord (1995) The Hamptons lecture: Our heritage – development and the millennium, *The Estates Gazette*, Issue 9501, 7 January, pp. 98–102.

Roberts, J. (1995a) Historic parks and gardens – listing, awareness and the future, *Journal of Architectural Conservation*, Vol. 1, No. 1, March, pp. 38–55.

Roberts, J. (1995b) Historic parks, gardens and urban spaces: how to recognise and what to conserve, *Context*, No. 45, March, pp. 27–8.

Scrase, T. (1991) Planning and archaeology – a case for full integration [1991] JPL 1103.

Stretton, A. and Bilikowski, K. (1993) Historic parks and gardens, *Conservation Bulletin*, Issue 21, November, pp. 4–5.

Wainwright, G. and Darvill, T. (1994) MARS measures monuments at risk, *Conservation Bulletin*, Issue 23, July, pp. 29–30.

Wheatley, C. (1995) Landlords to lobby EH on new listing proposals, *The Estates Gazettte*, 18 March, p. 42.

Appendix 1

SPAB – THE SOCIETY FOR THE PROTECTION OF ANCIENT BUILDINGS: MANIFESTO

Manifesto of William Morris, Founder, 1877

A Society coming before the public with such a name as that above written must needs explain how, and why, it proposes to protect those ancient buildings which, to most people doubtless, seem to have so many and such excellent protectors. This, then, is the explanation we offer.

No doubt within the last fifty years a new interest, almost like another sense, has arisen in these ancient monuments of art; and they have become the subject of one of the most interesting of studies, and of an enthusiasm, religious, historical, artistic, which is one of the undoubted gains of our time; yet we think that if the present treatment of them be continued, our descendants will find them useless for study and chilling to enthusiasm. We think that those last fifty years of knowledge and attention have done more for their destruction than all the foregoing centuries of revolution, violence, and contempt.

For Architecture, long decaying, died out, as a popular art at least, just as the knowledge of mediaeval art was born. So that the civilised world of the nineteenth century has no style of its own amidst its wide knowledge of the styles of other centuries. From this lack and this gain arose in men's minds the strange idea of the Restoration of ancient buildings; and a strange and most fatal idea, which by its very name implies that it is possible to strip from a building this, that, and the other part of its history – of its life that is – and then to stay the hand at some arbitrary point, and leave it still historical, living, and even as it once was.

In early times this kind of forgery was impossible, because knowledge failed the builders, or perhaps because instinct held them back. If repairs were needed, if ambition or piety pricked on to change, that change was of necessity wrought in the unmistakable fashion of the time; a church of the eleventh century might be added to or altered in the twelfth, thirteenth, fourteenth, fifteenth, sixteenth, or even the seventeenth or eighteenth centuries; but every change, whatever history it destroyed, left history in the gap, and was alive with the spirit of the deeds done midst its fashioning. The result of all this was often a building in which the many changes, though harsh and visible enough, were, by their very contrast, interesting and instructive and could by no possibility mislead. But those who make the changes wrought in our day under the name of Restoration, while professing to bring back a building to the best time of its history, have no guide but each his own individual whim to point out to them what is admirable and what contemptible; while the very nature of their tasks compels them to destroy something and to supply the gap by imagining what the earlier builders should or might have done. Moreover, in the course of this double process of destruction and addition the whole surface of the building is necessarily tampered with; so that the appearance of antiquity is taken away from such old parts of the fabric as are left, and there is no laying to rest in the spectator the suspicion of what may have been lost; and in short, a feeble and lifeless forgery is the final result of all the wasted labour.

It is sad to say, that in this manner most of the bigger Minsters, and a vast number of more humble buildings, both in England and on the Continent, have been dealt with by men of talent often, and worthy of better employment, but deaf to the claims of poetry and history in the highest sense of the words.

For what is left we plead before our architects themselves, before the official guardians of buildings, and before the public generally, and we pray them to remember how much is gone of the religion, thought and manners of time past, never by almost universal consent, to be Restored; and to consider whether it be possible to Restore those buildings, the living spirit of which, it cannot be too often repeated, was an inseparable part of that religion and thought, and those past manners. For our part we assure them fearlessly, that of all the Restorations yet undertaken the worst have meant the reckless stripping a building of some of its most interesting material features; while the best have their exact analogy in the Restoration of an old picture, where the partly perished work of the ancient craftsmaster has been made neat and smooth by the tricky hand of some unoriginal and thoughtless hack of today. If, for the rest, it be asked us to specify what kind of amount of art, style, or other interest in a building, makes it worth protecting, we answer, anything which can be looked on as artistic, picturesque, historical, antique, or substantial: any work, in short, over which educated, artistic people would think it worth while to argue at all.

It is for all these buildings, therefore, of all times and styles, that we plead, and call upon those who have to deal with them to put Protection in the place of Restoration, to stave off decay by daily care, to prop a perilous wall or mend a leaky roof by such means as are obviously meant for support or covering, and show no pretence of other art, and otherwise to resist all tampering with either the fabric or ornament of the building as it stands; if it has become inconvenient for its present use, to raise another building rather than alter or enlarge the old one; in fine to treat our ancient buildings as monuments of a bygone art, created by bygone manners, that modern art cannot meddle with without destroying.

Thus, and thus only, shall we escape the reproach of our learning being turned into a snare to us; thus, and thus only can we protect our ancient buildings, and hand them down instructive and venerable to those that come after us.

(Reproduced with the kind permission of the Society for the Protection of Ancient Buildings)

Appendix 2

ICOMOS – INTERNATIONAL COUNCIL OF MONUMENTS AND SITES: THE VENICE CHARTER

International Charter for the Conservation and Restoration of Monuments and Sites 1966

Imbued with a message from the past, the historic monuments of generations of people remain to the present day as living witnesses of their age-old traditions. People are becoming more and more conscious of the unity of human values and regard ancient monuments as a common heritage. The common responsibility to safeguard them for future generations is recognised. It is our duty to hand them on in the full richness of their authenticity.

It is essential that the principles guiding the preservation and restoration of ancient buildings should be agreed and be laid down on an international basis, with each country being responsible for applying the plan within the framework of its own culture and traditions.

By defining these basic principles for the first time, the Athens Charter of 1931 contributed towards the development of an extensive international movement which has assumed concrete form in national documents, in the work of ICOM and UNESCO and in the establishment by the latter of the International Centre for the Study of the Preservation and the Restoration of Cultural Property. Increasing awareness and critical study have been brought to bear on problems which have continually become more complex and varied; now the time has come to examine the Charter afresh in order to make a thorough study of the principles involved and to enlarge its scope in a new document.

Accordingly, the IInd International Congress of Architects and Technicians of Historic Monuments, which met in Venice from May 25th to 31st 1964, approved the following text:

Definitions

ARTICLE 1. The concept of an historic monument embraces not only the single architectural work but also the urban or rural setting in which is found the evidence of a particular civilisation, a significant development or an historic event. This applies not only to great works of art but also to more modest works of the past which have acquired cultural significance with the passing of time.

ARTICLE 2. The conservation and restoration of monuments must have recourse to all the sciences and techniques which can contribute to the study and safeguarding of the architectural heritage.

Aim

ARTICLE 3. The intention in conserving and restoring monuments is to safeguard them no less as works of art than as historical evidence.

Conservation

ARTICLE 4. It is essential to the conservation of monuments that they be maintained on a permanent basis.

ARTICLE 5. The conservation of monuments is always facilitated by making use of them for some socially useful purpose. Such use is therefore desirable but it must not change the lay-out or decoration of the building. It is within these limits only that modifications demanded by a change of function should be envisaged and may be permitted.

ARTICLE 6. The conservation of a monument implies preserving a setting which is not out of scale. Wherever the traditional setting exists, it must be kept. No new construction, demolition or modification which would alter the relations of mass and colour must be allowed.

ARTICLE 7. A monument is inseparable from the history to which it bears witness and from the setting in which it occurs. The moving of all or part of a monument cannot be allowed except where the safeguarding of that monument demands it or where it is justified by national or international interests of paramount importance.

ARTICLE 8. Items of sculpture, painting or decoration which form an integral part of a monument may only be removed from it if this is the sole means of ensuring their preservation.

Restoration

ARTICLE 9. The process of restoration is a highly specialised operation. Its aim is to preserve and reveal the aesthetic and historic value of the monument and is based on respect for original material and authentic documents. It must stop at the point where conjecture begins, and in this case moreover any extra work which is indispensable must be distinct from the architectural composition and must bear a contemporary stamp. The restoration in any case must be preceded and followed by an archaeological and historical study of the monument.

ARTICLE 10. Where traditional techniques prove inadequate, the consolidation of a monument can be achieved by the use of any modern technique for conservation and construction, the efficacy of which has been shown by scientific data and proved by experience.

ARTICLE 11. The valid contributions of all periods to the building of a monument must be respected, since unity of style is not the aim of a restoration. When a building includes the superimposed work of different periods, the revealing of the underlying state can only be justified in exceptional circumstances and when what is removed is of little interest and the material which is brought to light is of great historical, archaeological or aesthetic value, and its state of preservation good enough to justify the action. Evaluation of the importance of the elements involved and the decision as to what may be destroyed cannot rest solely on the individual in charge of the work.

ARTICLE 12. Replacements of missing parts must integrate harmoniously with the whole, but at the same time must be distiguishable from the original so that restoration does not falsify the artistic or historic evidence.

ARTICLE 13. Additions cannot be allowed except in so far as they do not detract from the interesting parts of the building, its traditional setting, the balance of its composition and its relation with its surroundings.

Historic sites

ARTICLE 14. The sites of monuments must be the object of special care in order to safeguard their integrity and ensure that they are cleared and presented in a seemly manner. The work of conservation and restoration carried out in such places should be inspired by the principles set forth in the foregoing articles.

Excavations

ARTICLE 15. Excavations should be carried out in accordance with scientific standards and the recommendation defining international principles to be applied in the case of archaeological excavation adopted by UNESCO in 1956.

Ruins must be maintained and measures necessary for the permanent conservation and protection of architectural features and of objects discovered must be taken. Furthermore, every means must be taken to facilitate the understanding of the monument and to reveal it without ever distorting its meaning.

All reconstruction work should however be ruled out *a priori*. Only anastylosis, that is to say, the reassembling of existing but dismembered parts, can be permitted. The material used for integration should always be recognisable and its use should be the least that will ensure the conservation of a monument and the reinstatement of its form.

Publication

ARTICLE 16. In all works of preservation, restoration or excavation, there should always be precise documentation in the form of analytical and critical reports, illustrated with drawings and photographs.

Every stage of the work of clearing, consolidation, rearrangement and integration, as well as technical and formal features indentified during the course of the work, should be included. This record should be placed in the archives of a public institution and made available to research workers. It is recommended that the report should be published.

The following persons took part in the work of the Committee for drafting the International Charter for the Conservation and Restoration of Monuments:

Mr PIERO GAZZOLA (Italy), Chairman
Mr RAYMOND LEMAIRE (Belgium), Reporter
Mr JOSE BASSEGODA-NONELL (Spain)
Mr LUIS BENAVENTE (Portugal)
Mr DJURDJE BOSKOVIC (Yugoslavia)
Mr HIROSHI DAIFUKU (UNESCO)
Mr P.L. DE VRIEZE (Netherlands)
Mr HARALD LANGBERG (Denmark)
Mr MARIO MATTEUCCI (Italy)
Mr JEAN MERLET (France)
Mr CARLOS FLORES MARINI (Mexico)
Mr ROBERTO PANE (Italy)
Mr S.C.J. PAVEL (Czechoslovakia)
Mr PAUL PHILIPPOT (International Centre for the Study of the Preservation and Restoration of Cultural Property)
Mr VICTOR PIMENTEL (Peru)
Mr HAROLD PLENDERLEITH (International Centre for the Study of the Preservation and Restoration of Cultural Property)
Mr DEOCLECIO REDIG DE CAMPOS (Vatican)
Mr JEAN SONNIER (France)
Mr FRANCIOS SORLIN (France)
Mr EUSTATHIOS STIKAS (Greece)
Mrs GERTRUD TRIPP (Austria)
Mr JAN ZACHWATOVICZ (Poland)
Mr MUSTAFA S. ZBISS (Tunisia)

Appendix 3

LIST OF CHARTERS

Athens Charter – 1931, the foundation of conservation.

UNESCO – 1954, The Hague Convention on the protection of cultural properties in case of armed conflicts.

ICOMOS – 1964 (the foundation of ICOMOS), Venice Charter, the basis of all modern conservation.

Organisation of American States – 1967, Quito Recommendations.

UNESCO – 1972 The World Heritage Convention.

The Council of Europe – 1975, European Charter of the Architectural Heritage (Amsterdam).

ICOMOS – 1976, Charter of Cultural Tourism.

ICOMOS – 1978, Burra Charter, adopted by Australia ICOMOS as a supplement to the Venice Charter.

ICOMOS – 1981, Florence Charter, relating to the preservation of historic gardens.

ICOMOS – 1982, Declaration of Dresden, Reconstruction of Monuments Destroyed by War.

ICOMOS – 1982, Charter for the Preservation of Quebec's Heritage Deschambault Declaration (Quebec).

ICOMOS – 1983, The Appleton Charter on the Protection and Enhancement of the Built Environment (Canada).

The Council of Europe – 1985, Convention for the Protection of the Architectural Heritage (Granada).

ICOMOS – 1987, Charter of Petropolis, Brazil, for the conservation of historic towns and urban areas.

ICOMOS – 1987, Charter for the conservation of historic towns and urban areas (Washington).

ICOMOS – 1989, Charter for archaeological heritage management (Lausanne)

ICOMOS – 1992, Charter for the Conservation of Places of Cultural Heritage Value (New Zealand).

Appendix 4

DIRECTORY OF ADDRESSES

Ancient Monuments Society, St Anns Vestry Hall, 2 Church Entry, London, EC4V 5HB. Tel. 0171 236 3934, 0171 329 3677.

Antique & Decorative Lighting Dealers Association, Littleton House, Littleton, Somerset. TA11 6NP. Tel. 01458 72341, 01458 733352.

Architectural Heritage Fund, 27 John Adam Street, London, WC2N 6HZ. Tel. 0171 925 0199.

The Architectural Salvage Index, c/o Hutton & Rostron, Netley House, Gomshall, Surrev, GU5 9QA. Tel. 01483 203221.

Association for Industrial Archaeology, Ironbridge Gorge Museum. The Wharfage, Ironbridge. Telford, Shropshire. TF8 7AW. Tel. 01952 432751.

Association for Studies in the Conservation of Historic Buildings, Institute of Archaeology, 31–34 Gordon Square, London, WC1H 0PY.

Association of Bronze and Brass Founders, Heathcote House, 136 Hagley Road, Edgbaston, Birmingham, West Midlands, B16 9PN. Tel. 0121 454 4141.

Association of Conservation Officers, 24 Middle Street, Stroud, Gloucestershire, GL5 1DZ. Tel. 01453 753949/753358

Association of Preservation Trusts, 27 John Adam Street, London, WC2N 6HZ. Tel. 0171 930 1629.

The Baptist Building Fund, 2 Highlands Road, Long Ashton, Bristol, BS18 9EN.

Brick Development Association, Woodside House, Winkfield, Windsor, Berkshire, SL4 2DX. Tel. 01344 885651.

British Aggregate Construction Materials Industries, 156 Buckingham Palace Road, London, SWI W9TR. Tel. 0171 7308194.

British Antique Furniture Restorers Association. The Old School, Longmor, Shrewsbury, Shropshire, SW55 7PP. Tel. 01743 718162.

British Artist Blacksmiths Association, Lyndhurst, Carlton Husthwaite. Thirsk, North Yorkshire, YO7 2BJ. Tel. 01845 501415.

British Glass Manufacturers Confederation, Northumberland Road, Sheffield, South Yorkshire, S10 2UA. Tel. 0114 286 1073.

The British Institute of Non-Destructive Testing, 1 Spencer Parade, Northampton, Northamptonshire, NN1 5AA. Tel. 01604 30124.

British Lime Association, 156 Buckingham Palace Road. London, SW1 W9TF. Tel. 0171 730 8194.

British Society of Master Glass Painters – Conservation Committee, c/o Sarah Brown, RCHME, Fortress House, 23 Savile Row, London, W1X 2JQ. Tel. 0171 973 3091.

British Standards Institution, 389 Chiswick High Road, London, WW4 4AL. Tel. 0181 996 9000.

British Wood Preserving and Damp-proofing Association, Building No. 6. The Office Village, 4 Romford Road, Stratford, London, E15 4EA. Tel. 0181 519 2588.

The Building Crafts & Conservation Trust, Kings Gate, Dover Castle, Dover, Kent, CT16 1HU. Tel. 01304 225066.

Building Employers Confederation, 82 New Cavendish Street, London, W1M 8AD. Tel. 0171 580 5588.

Building Research Establishment Advisory Service, Bucknalls Lane, Garston, Watford, Hertfordshire, WD2 7JR. Tel. 01923 664664.

CAPCIS – Corrosion Control, Materials Conservation & Environmental Analysis, Bainbridge House, Granby Row, Manchester, M1 2PW. Tel. 0161 236 6573.

Capel – Chapels Heritage Society, c/o West Glamorgan Archive Service, County Hall, Oystermouth Road, Swansea, West Glamorgan, Wales, SA1 3SN. Tel. 01792 471589.

Cathedral Architects Association, Harcourt Offices, Hemingford Grey, Huntingdon, Cambridgeshire, PE18 9BJ. Tel. 01480 461101.

The Cathedrals Fabric Commission for England, 83 London Wall, London, EC2M 5NA. Tel. 0171 638 0971.

Centre for Conservation Studies, Institute of Advanced Architectural Studies, University of York. The King's Manor, York, YO1 2EP. Tel. 01904 433963.

The Chapels Society, Rookery Farmhouse, Laxfield, Nr Woodbridge, Suffolk, IP13 8JA. Tel. 01986 798308.

Chartered Institute of Building, Englemere, Kings Ride, Ascot, Berkshire, SL5 8BJ. Tel. 01344 23355.

The Church Monuments Society, c/o The Royal Armouries, H.M. Tower of London, London, EC3N 4AB.

The Churches Conservation Trust (formerly The Redundant Churches Fund), 98 Fleet Street, London, EC4Y 1DH. Tel. 0171 936 2285.

The Civic Trust, 17 Carlton House Terrace, London, SW1Y 5AW. Tel. 0171 930 0914.

The College of Masons, 42 Magdalen Road, Wandsworth, London, SW18 3NP. Tel 0181 874 8363.

COTAC – Conference on Training in Architectural Conservation, Keysign House, 429 Oxford street, London, W1R 2HD. Tel. 0171 973 3615.

The Conservation Unit, Museums & Galleries Commission, 16 Queen Anne's Gate, London, SW1H 9AA. Tel. 0171 233 4200.

Construction History Society, c/o Chartered Institute of Building, Englemere, Kings Ride, Ascot, Berkshire, SL5 8BJ. Tel. 01344 23355.

Copper Development Association, Orchard House, Mutton Lane, Potters Bar, Hertfordshire, EN6 3AP. Tel. 01707 650711.

Council for British Archaeology, Bowes Morrell House, 111 Walmgate, York, North Yorkshire, YO1 2UA. Tel. 01904 671417.

The Council for thc Care of Churches, 83 London Wall, London, EC2M 5NA. Tel. 0171 638 0971.

Council of Europe, Division of Cultural Heritage, F-67006 Strasbourg Cedex, France.

Department of National Heritage, 2–4 Cockspur Street, London SW1Y 5DH. Tel. 0171 211 6210.

Department of the Environment, 2 Marsham Street, London SW1P 3EB. Tel. 0171 276 0900

DOCOMOMO UK. The Building Centre, 26 Store Street, London, WC1E 7BT. Tel. 0171 637 0276.

Dry Stone Walling Association of Great Britain, c/o YFC Centre, National Agricultural Centre, Stoneleigh Park, Kenilworth, Warwickshire, CV8 2LG. Tel. 0121 378 0493.

Ecclesiastical Architects and Surveyors Association, Scan House, 29 Radnor Cliff, Folkestone, Kent, CT20 2JJ. Tel. 01227 459401.

English Heritage, Fortress House, 23 Savile Row, London, W1X 1AB. Tel. 0171 973 3000.

Europa Nostra, 86 Vincent Square, London, SW1P 2PG.

Federation of Master Builders, Gordon Fisher House, 14-15 Great James Street, London, WC1N 3DP. Tel. 0171 242 7583.

Fire Protection Association, 140 Aldergate Street, London, EC1A 4HX. Tel. 0171 606 3757.

Friends of Friendless Churches, St Ann's Vestry Hall, 2 Church Entry, London, EC4V 5HB. Tel. 0171 236 3934.

The Furniture History Society, 1 Mercedes Cottages, St John's Road, Haywards Heath, West Sussex, RH16 6EH. Tel. 01444 413845.

Garden History Society, Station House, Church Lane, Wickwar, Gloucestershire, GL12 8NB. Tel. 01454 294888.

The Georgian Group, 37 Spital Square, London, E1 6DY. Tel. 0171 377 1722.

The Glass and Glazing Federation, 44–48 Borough High Street, London, SE1 1XB. Tel. 0171 403 7177.

The Guild of Architectural Ironmongers, 8 Stepney Green, London, E1 3JU. Tel. 0171 790 3431.

The Heritage Building Contractors Group (UK), c/o Linford Group Ltd, Quonians, Lichfield, Staffordshire, WS13 7LB. Tel. 01543 414234.

Heritage Lottery Fund. The National Heritage Memorial Fund, 10 St James's Street, London, SW1A 1EF. Tel. 0171 649 1345.

Historic Chapels Trust, 29 Thurloe Street, London, SW7 2LQ. Tel. 0171 584 6072.

Historic Churches Preservation Trust, Fulham Palace, London, SW6 6EA. Tel. 0171 736 3054

Historic Farm Buildings Group, c/o Museum of English Rural Life, University of Reading, Whiteknights, PO Box 229, Reading, Berkshire, RG6 2AG. Tel. 01734 318663.

Historic Houses Association, 2 Chester Street, London, SW1X 7BB. Tel. 0171 259 5688.

ICCROM, 13 Via de San Michele, Rome, Italy. Tel. 01039 6 587 901.

Institute of Field Archaeologists, Metallurgy & Materials Building, University of Birmingham, Edgbaston, Birmingham, West Midlands, B15 2TT. Tel. 0121 471 2788.

Institute of Quarrying, 7 Regent Street, Nottingham, Nottinghamshire, NG1 5BS. Tel. 0115 941 1315.

Institution of Civil Engineers, 1–7 Great George Street, London, SW1P 3AA. Tel. 0171 222 7722.

Institution of Structural Engineers, 11 Upper Belgrave Street, London, SW1X 8BH. Tel. 0171 235 4535.

Interior Decorators and Designers Association, Crest House, 102–104 Church Road, Teddington, Middlesex, TW11 8PY. Tel. 0181 977 1105.

ICOMOS UK, 10 Barley Mow Passage, London, W4 4PH. Tel. 0181 994 6477.

The Institutional Institute for Conservation of Historic and Artistic Works, 6 Buckingham Street, London, WC2N 6BA. Tel. 0171 839 5975.

The Landmark Trust, Shottesbrooke, Maidenhead, Berkshire, SL6 3SW. Tel. 01628 825925.

Lead Sheet Association, St Johns Road, Tunbridge Wells, Kent, TN4 9XA. Tel. 01892 513351.

LOTS – Living Over The Shop, c/o Ann Petherick, Institute of Advanced

Architectural Studies, University of York, The King's Manor, York, YO1 2EP. Tel. 01904 433963.

Master Carvers Association, Unit 20, 21 Wren Street, London, WC1X 0HF. Tel. 0171 278 8759.

The Methodist Church, Property Division, Central Buildings, Oldham Street, Manchester, M1 1JQ. Tel. 0161 236 5194.

The Millennium Commission, 2 Little Street, London, SW1P 3DH. Tel. 0171 340 2001.

National Association of Decorative and Fine Art Societies, 8a Lower Grosvenor Place, London, SW1W 0EN. Tel. 0171 340 2001.

National Council of Master Thatchers Association, Thatcher's Rest, Levens Green, Great Mundon, Nr Ware, Hertfordshire, SG11 1HD. Tel. 01920 438710.

National Heritage Memorial Fund, 10 St James's Street, London, SW1A 1EF. Tel. 0171 930 0963.

National Monument Record Centre, Kemble Drive, Swindon, SN2 2GZ. Tel. 01793 414600.

National Society of Master Thatchers, The Castle, Great Bedwyn, Marlborough, Wiltshire, SN8 3LU. Tel. 01672 870225.

The National Trust, 36 Queen Anne's Gate, London, SW1H 9AS. Tel. 0171 222 9251.

Natural Slate Quarries Association, 26 Store Street, London, WC1E 7BT. Tel. 0171 323 3770.

The Orton Trust, PO Box 34, Rothwell, Kettering, Northamptonshire, NN14 6XP. Tel. 01536 710692.

Public Record Office, Chancery Lane, London, WC2A 1LR. Tel. 0181 876 3444.

Railway Heritage Trust, Melton House, 65–67 Clarendon Road, Watford, Hertfordshire, WD1 1DP. Tel. 01923 240250.

Royal Archaeological Institute, c/o Society of Antiquaries of London, Burlington House, Piccadilly, London, W1Y 0HS

Royal Commission on the Historical Monuments of England, National Monument Record Centre, Kemble Drive, Swindon, SN2 2GZ. Tel. 01793 414600.

Royal Fine Art Commission, 7 St James's Square, London, SW1Y 4JU. Tel. 0171 839 6537.

Royal Institute of British Architects, 66 Portland Place, London, W1N 4AD. Tel. 0171 580 5533.

Royal Institution of Chartered Surveyors (Building Conservation Group, Building Surveyors Division), 12 Great George Street, London, SW1P 3AE. Tel. 0171 222 7000.

Royal Town Planning Institute, 26 Portland Place, London, W1N 4BE. Tel. 0171 636 9107.

Rural Development Commission, 141 Castle Street, Salisbury, Wiltshire, SP1 3TP. Tel. 01722 336255.

SAVE Britain's Heritage, 68 Battersea High Street, London, SW11 3HX. Tel. 0171 228 3336.

Society of Architectural Historians of Great Britain, 23b Home Park Road, Wimbledon Park, London, SW19 7HP.

Society for the Protection of Ancient Buildings, 37 Spital Square, London, E1 6DY. Tel. 0171 377 1644.

The Stained Glass Museum, North Triforium, Ely Cathedral, c/o Mrs S. Mathews, 10 Ferry Lane, Chesterton, Cambridge, CB4 1NT. Tel. 01223 327367.

Stone Federation of Great Britain, 18 Mansfield Street, London, W1M 9FG. Tel. 0171 580 5404.

The Temple Trust, c/o Gunnersbury Park Museum, Gunnersbury Park, Popes Lane, London, W3 8LQ. Tel. 0181 992 2248.

The Theatres Trust, 22 Charing Cross Road, London, WC2H 0HR. Tel. 0171 836 8591.

Tiles and Architectural Ceramics Society, c/o Decorative Art Department, Liverpool Museum, William Brown Street, Liverpool, L3 8EN. Tel. 0151 207 0001.

Timber Research and Development Association, Stocking Lane, Hughenden Valley, High Wycombe, Buckinghamshire, HP14 4ND. Tel. 01494 563091.

The Twentieth Century Society, 70 Cowcross Street, London, EC1M 6BP. Tel. 0171 250 3857.

United Kingdom Institute for Conservation of Historic and Artistic Works, 6 Whitehorse Mews, Westminster Bridge Road, London, SE1 7QD. Tel. 0171 620 3371.

Upkeep (formerly The Building Conservation Trust), Apartment 39, Hampton Court Palace, East Molesey, Surrey, KT8 9BS. Tel. 0181 943 2277.

Vernacular Architecture Group, 16 Falna Crescent, Coton Green. Tarnworth, Staffordshire, B79 8JS. Tel. 01827 69434.

The Victorian Society, 1 Priory Gardens, Bedford Park, London, W4 1TT. Tel. 0181 994 1019.

Victoria & Albert Museum, London, SW7 2RL. Tel. 0171 938 8500

Wallpaper History Society, c/o Archives, Arthur Sanderson & Sons Ltd, 100 Acres, Oxford Road, Uxbridge, Greater London, UB8 1HY. Tel. 01895 238244.

Weald & Downland Open Air Museum, Singleton, Chichester, Sussex, PO18 0EU. Tel. 0124 363 348

The Worshipful Company of Glaziers and Painters of Glass, Glaziers Hall, 9 Montague Close, London, SE1 9DD. Tel. 0171 403 3300.

Appendix 5

ICOMOS – INTERNATIONAL COUNCIL ON MONUMENTS AND SITES: THE LAUSANNE CHARTER

Charter for the Protection and Management of the Archaeological Heritage 1989

It is widely recognised that a knowledge and understanding of the origins and development of human societies is of fundamental importance to humanity in identifying its cultural and social roots.

The archaeological heritage constitutes the basic record of past human activities. Its protection and proper management is therefore essential to enable archaeologists and other scholars to study and interpret it on behalf of and for the benefit of present and future generations.

The protection of this heritage cannot be based upon the application of archaeological techniques alone. It requires a wider basis of professional and scientific knowledge and skills. Some elements of the archaeological heritage are components of architectural structures and in such cases must be protected in accordance with the criteria for the protection of such structures laid down in the 1966 Venice Charter on the Conservation and Restoration of Monuments and Sites. Other elements of the archaeological heritage constitute part of the living traditions of indigenous peoples, and for such sites and monuments the participation of local cultural groups is essential for their protection and preservation.

For these and other reasons the protection of the archaeological heritage must be based upon effective collaboration between professionals from many disciplines. It also requires the cooperation of government authorities, academic researchers, private *or* public enterprise, and the general public. This Charter therefore lays down principles relating to the different aspects of archaeological heritage management. These include the responsibilities of public authorities and legislators, principles relating to the professional performance of the processes of inventorisation, survey, excavation, documentation, research, maintenance, conservation, preservation, reconstruction, information, presentation, public access and use of the heritage, and the qualification of professionals involved in the protection of the archaeological heritage.

The charter has been inspired by the success of the Venice Charter as guidelines and source of ideas for policies and practice of governments as well as scholars and professionals.

The charter has to reflect very basic principles and guidelines with global validity. For this reason it cannot take into account the specific problems and possibilities of regions or countries. The charter should therefore be supplemented at regional and national level by further principles and guidelines for these needs.

Article 1: Definition and introduction

The 'archaeological heritage' is that part of the material heritage in respect of which archaeological methods provide primary information. It comprises all vestiges of human existence and consists of places relating to all manifestations of human activity, abandoned structures, and remains of all kinds (including subterranean and underwater sites). Together with all the portable cultural material associated with them.

Article 2: Integrated protection policies

The archaeological heritage is a fragile and non-renewable cultural resource. Land use must therefore be controlled and developed in order to minimise the destruction of the archaeological heritage.

Policies for the protection of the archaeological heritage should constitute an integral component of policies relating to land use, development, and planning as well as of cultural environmental and educational policies. *The policies for the protection of the archaeological heritage should be kept under continual review, so that they stay up to date.* The creation of archaeological reserves should form part of such policies.

The protection of the archaeological heritage should be integrated into planning policies at international, national, regional and local level.

Active participation by the general public must form part of policies for the protection of the archaeological heritage. This is essential where the heritage of indigenous peoples is involved. Participation must be based upon access to the knowledge necessary for decision-making. The provision of information to the general public is therefore an important element in integrated protection.

Article 3: Legislation and economy

The protection of the archaeological heritage should be considered as a moral obligation upon all human beings; it is also a collective public responsibility. This obligation must be acknowledged through relevant legislation and the provision of adequate funds for the supporting programmes necessary for effective heritage management.

The archaeological heritage is common to all human society and it should therefore be the duty of every country to ensure that adequate funds are available for its protection.

Legislation should afford protection to the archaeological heritage that is appropriate to the needs, history, and traditions of each country and region, providing for *in situ* protection and research needs.

Legislation should be based on the concept of the archaeological heritage as the heritage of all humanity and of groups of peoples, and not restricted to any individual person or nation.

Legislation should forbid the destruction, degradation or alteration through changes of any archaeological site or monument or to their surroundings without the consent of the relevant archaeological authority.

Legislation should in principle require full archaeological investigation and documentation in cases where the destruction of the archaeological heritage is authorised.

Legislation should require, and make provision for. The proper maintenance, *management* and conservation of the archaeological heritage.

Adequate legal sanctions should be prescribed in respect of violations of archaeological heritage legislation.

If legislation affords protection only to those elements of the archaeological heritage which are registered in a selective statutory inventory, provision should be made for the temporary protection of unprotected or newly discovered sites and monuments until an archaeological evaluation can be carried out.

Development projects constitute one of the greatest physical threats to the archaeological heritage. A duty for developers to ensure that archaeological heritage impact studies are carried out before development schemes are implemented, should therefore be embodied in appropriate legislation, with a stipulation that the costs of such studies are to be included in project costs. The principle should also be established in legislation that development schemes should be designed in such a way as to minimise their impact upon the archaeological heritage.

Article 4: Survey

The protection of the archaeological heritage must be based upon the fullest possible knowledge of its extent and nature. General survey of archaeological resource is therefore an essential working tool in developing strategies for the protection of the archaeological heritage. Consequently archaeological survey should be a basic obligation in the protection and management of the archaeological heritage.

At the same time, inventories constitute primary resource databases for scientific study and research. The compilation of inventories should therefore be regarded as a continuous, dynamic process. It follows that inventories should comprise information at various levels of significance and reliability, since even superficial knowledge can form the starting point for protectional measures.

Article 5: Investigation

Archaeological knowledge is based principally on the scientific investigation of the archaeological heritage. Such investigation embraces the whole range of methods from non-destructive techniques through sampling to total excavation.

It must be an overriding principle that the gathering of information about the archaeological heritage should not destroy any more archaeological evidence than is necessary for the protectional or scientific objectives of the investigation. Non-destructive techniques, aerial and ground survey, and sampling should therefore be encouraged wherever possible, in preference to total excavation.

As excavation always implies the necessity of making a selection of evidence to be documented and preserved at the cost of losing other information and possibly even the total destruction of the monument, a decision to excavate should only be taken after thorough consideration.

Excavation should be carried out on sites and monuments threatened by development, land-use change, looting, or natural deterioration.

In exceptional cases, unthreatened sites may be excavated to elucidate research problems or to interpret them more effectively for the purpose of presenting them to the public. In such cases excavation must be preceded by thorough scientific evaluation of the significance of the site Excavation should be partial, leaving a portion undisturbed for future research.

A report conforming to an agreed standard should be made available to the scientific community and should be incorporated in the relevant inventory within a reasonable period after the conclusion of the excavation.

Excavations should be conducted in accordance with the principles embodied in the 1956 UNESCO Recommendations on International Principles Applicable to Archaeological Excavations and with agreed international and national professional standards.

Article 6: Maintenance and conservation

The overall objective of archaeological heritage management should be the preservation of monuments and sites *in situ, including proper long-term conservation and curation of all related records and collections, etc.* Any transfer of elements of the heritage to new locations represents a violation of the principle of preserving the heritage in its original context. This principle stresses the need for proper maintenance, conservation and management. It also asserts the principle that the archaeological heritage should not be exposed by excavation or left exposed after excavation if provision for its proper maintenance and management after excavation cannot be guaranteed.

Local commitment and participation should be actively sought and encouraged as a means of promoting the maintenance of the archaeological heritage. This principle is especially important when dealing with the heritage of indigenous peoples or local cultural groups. In some cases it may be appropriate to entrust responsibility for the protection and management of sites and monuments to indigenous peoples.

Owing to the inevitable limitations of available resources, active maintenance will have to be carried out on a selective basis. It should therefore be applied to a sample of the diversity of sites and monuments, based upon a scientific assessment of their significance and representative character, and not confined to the more notable and visually attractive monuments.

The relevant principles of the 1956 UESCO Recommendations should be applied in respect of the maintenance and conservation of the archaeological heritage.

Article 7: Presentation, information, reconstruction

The presentation of the archaeological heritage to the general public is an essential method of promoting an understanding of the origins and development of modern societies. At the same time it is the most important means of promoting an understanding of the need for its protection.

Presentation and information should be conceived as a popular interpretation of the current state of knowledge, and it must therefore be revised frequently. It should take account of the multi-faceted approaches to an understanding of the past.

Reconstructions serve two important functions: experimental research and interpretation. They should, however, be carried out with great caution, so as to avoid disturbing any surviving archaeological evidence, and they should take account of evidence from all sources in order to achieve authenticity. Where possible and appropriate, reconstructions should not be built immediately on the archaeological remains, and should be identifiable as such.

Article 8: Professional qualifications

High academic standards in many different disciplines are essential in the management of the archaeological heritage. The training of an adequate number of qualified professionals in the relevant fields of expertise should therefore be an important objective for the educational policies in every country. The need to develop expertise in certain highly specialised fields calls for international cooperation. *Standards of professional training and professional conduct should be established and maintained.*

The objective of academic archaeological training should take account of the shift in conservation policies from excavation to *in situ* preservation. It should also take into account the fact that the study of the history of indigenous peoples is as important in preserving and understanding the archaeological heritage as the study of outstanding monuments and sites.

The protection of the archaeological heritage is a process of continuous dynamic development. Time should therefore be made available to professionals working in this field to enable them to update their knowledge. Postgraduate training programmes should be developed with special emphasis on the protection and management of the archaeological heritage.

Article 9: International cooperation

The archaeological heritage is the common heritage of all humanity. International cooperation is therefore essential in developing and maintaining standards in its management.

There is an urgent need to create international mechanisms for the exchange of information and experience among professionals dealing with archaeological heritage management. This requires the organisation of conferences, seminars, workshops, etc. at global as well as regional level, and the establishment of regional centres for postgraduate studies. ICOMOS, through its specialised groups, should promote this aspect in its medium and long-term planning.

International exchanges of professional staff should also be developed as a means of raising standards of archaeological heritage management.

Technical assistance programmes in the field of archaeological heritage management should be developed under the auspices of ICOMOS.

Appendix 6

NEW CONTROLS OVER BUILDINGS IN CONSERVATION AREAS

On 3rd June 1995 the Town and Country Planning General Development Order 1988 was replaced by the Town and Country Planning (General Development Procedure) Order 1995 (SI 1995/419) (hereafter referred to as the GDPO) and the Town and Country Planning (General Permitted Development) Order 1995 (SI 1995/418) (hereafter referred to as the GPDO).

The new controls

Article 4(2) of the GPDO allows district or county planning authorities to make a direction to bring planning control over certain permitted development items within conservation areas (in relation to a building or land *which fronts a relevant location*: a highway, waterway, or open space). This provides wider opportunities for planning control than previously existed under article 4 of the GDO 1988 (see pp. 219–20 and 236) and, of significance, does not require the prior approval of the Secretary of State for the Environment. Article 4(5) specifies the issues indicated in Schedule 2 of the GPDO which may be brought under control within a conservation area. These were highlighted (with associated new procedures for article 4(2) directions) in DoE Circular 9/95 and include the following items within Schedule 2 of the GPD: Part 1 – Classes A, C, D, E, F, H; Part 2 – Classes A and C; and Part 31 (where the issue in question would front a relevant location). The following points should be noted:

1. Under Class A Part 1 of Schedule 2 to the GPDO development not permitted under A.1(d) includes the insertion, enlargement, improvement or other alteration of a window in an existing wall of a dwellinghouse if the part of the building so affected is within 2 metres of the boundary of its curtilage and would not exceed 4 metres in height.

   While there have been instances of planning appeals determining that window alterations fall within the definition of development under s. 55 of the Town and Country Planning Act 1990 (being an 'other operation' which materially affects the external appearance of a building) the amendment of this provision via the GDPO brings more clarity to the issue of whether window alterations can be classed as development and brought under planning control. It should be noted, however, that other features (such as doors) which 'materially affect the external appearance of a building' may be classed as 'development'.

The difficulty in determining whether external alterations 'materially affect the external appearance' was considered in the recent case of *Burroughs Day v. Bristol, City Council* [1996] EGCS 10. Although the case concerned alterations to a listed building, the court highlighted the fact that the issue would equally apply to an unlisted building within a conservation area. In relation to window replacements located within the roof to the front elevation, it was considered that for the external appearance to be 'materially' affected would depend, in large part, on the degree of 'visibility'. In this case the court found that the window replacements did not constitute development. This was due to the fact that the roof alterations would not be visible from any street or from any window of any building nearby except from the top two floors of one office building and from the air.

2. Under Part 1 of Schedule 2 to the GPDO permitted development rights for the erection, alteration or removal of a chimney on a dwellinghouse or on a building within the curtilage of a dwellinghouse may be removed via an article 4(2) direction. This is not dependent on the criterion of 'fronting a relevant location'.
3. Under Class Part 31 of Schedule 2 to the GPDO permitted development rights for the demolition of a gate, fence, wall or other means enclosure within the curtilage of a dwellinghouse may be removed. This follows the replacement of the 1994 demolition direction with the Town and Country Planning (Demolition – Description of Buildings) Direction 1995. Under the 1995 direction such demolition is development only if it is within a conservation area (see DoE Circular 10/95).

The new procedures

The new discretionary power available to local planning authorities to make an article 4(2) direction (without the consent of the Secretary of State) is governed by strict procedures which means that residents who will be affected by a proposed direction must be notified about the proposal and their views considered before a direction is implemented. Article 6 of the GPDO provides that:

1. Notice must be given of an article 4(2) direction as soon as practicable after it has been made by:
   (i) a local advertisement
   (ii) serving a notice on the owner or occupier of every dwellinghouse (except where this is impracticable)

2. The notice must:
   (i) provide a description of the development affected by the direction
   (ii) indicate the conservation area (or part of an area) to which the direction relates
   (iii) indicate the overall effect of the direction
   (iv) name a place where the direction (and accompanying map) may be inspected
   (v) specify a period of at least 21 days within which representations may be made
   (vi) explain that the direction is made under article 4(2) of the GPDO
3. The local planning authority must:
   (i) consider any representations to the notices that have been received within the specified period
   (ii) confirm the direction only after a minimum of 28 days has elapsed from the date the last notice was served or published and not later than six months after the direction was made
   (iii) provide a further notice (by local advertisement and directly to owners and occupiers affected) when a direction has been confirmed
4. District (except metropolitan districts) and county planning authorities must notify each other when an article 4(2) is made by either authority relating to land within an area administratively covered by the other authority.

Appendix 7

INFORMATION UPDATE

The following information updates the text in Chapters 1 to 9 to February 1996. Further information sources are also provided.

Chapter 1

pp. 1–15 Definition of a listed building

In *Kennedy v. Secretary of State for Wales* [1996] EGCS 17 it was confirmed that a 'clarion clock' which 'rested on the floor by its own colossal weight' formed part of a listed building. The fact that it was free standing was not conclusive that it was not a fixture.

Further clarification of the term *curtilage* and its application to listed buildings may be found in an appeal case which considered whether the main entrance gates to Kentwell Hall, Long Melford were within the curtilage of the listed building (APP/F/90/D3505/000001 [1995] JPL 987).

p. 27 Consultations on listing proposals

Consultations were first carried out in early 1995 regarding proposals to list 40 commercial, railway and industrial buildings. In November 1995 it was announced that 13 commercial, 3 industrial and 5 railway buildings spread across the country had been added to the statutory list (DNH News Release 218/95).

In a further News release (133/95) it was announced that four more buildings would be subject to a similar consultation process:

- Richmond Swimming Baths, Old Deer Park Road, Richmond-upon-Thames (1964–66: Proposed Grade II)
- Gillet House, Chichester Theological college (1963–65: Proposed Grade II*)
- St Mary's School, Leasowe Road, Wallasey, Merseyside (1962: Proposed Grade II*)

- The former chapel at Hopwood Hall college, Middleton, Greater Manchester (1964–65: Proposed Grade II)

The Secretary of State indicated that immediate spot-listing would be considered if there is any threat of pre-emptive action to any of these buildings before the consultation process has been completed. It was further indicated that consultations on all proposed listings would not be possible until a provisional listing system, during the consultation period, has been put in place. This is to be considered in the forthcoming *Heritage Green Paper* which is expected to be published in the spring of 1996.

p. 27 Number and grade of listed buildings

The figures provided in Planning Policy Guidance Note 15 on p. 27 are approximate. The total number of listed building in England as recorded at 31 December 1995 was 448,916 of which 6,081 were listed grade I. No exact figures have been recorded to differentiate between the number of grade II* and grade II buildings (the total of both grades being 442,835).

Chapter 2

pp. 35–37 Demolition or alteration

Following the case of *Shimizu(UK) Ltd v. Westminster City Council* [1995] 23 EG 118, the need for clarification on what works constitute demolition or alteration is even greater. As the directions on the handling of listed building consent applications require, in the case of demolition applications, notification to the RCHME, National Amenity Societies and (in some cases) English Heritage, the proposed new procedures are likely to consider this aspect when they are eventually formulated (see below).

Chapter 3

pp. 78–81 Procedures for the handling of listed building consent applications

The first draft of PPG 15 was subject to a consultation period before the guidance was finally published in September 1994. During this period proposals were made to change the procedures for the handling of listed building consent applications. These were not brought through to the final published guidance and the directions contained in DoE Circular 8/87 remained in force. However, it is proposed that new direction procedures will be announced in the summer of 1996. These will be contained in a DoE Circular.

p. 82 Article 4 directions

The new GPDO, as well as creating the new article 4(2) direction procedure in relation to conservation areas (see Appendix 6), has also created a new article 4(1) direction procedure in relation to listed buildings. The procedure has not significantly changed in this context. By article 5(3) of the GPDO 1995 approval of the Secretary of State for an article 4(1) direction is not needed in the case of a listed building, a building which is notified to the LPA by the Secretary of State as a building of special architectural or historic interest, or for development within the curtilage of a listed building. The only exception to this is where it involves the carrying out of development by a statutory undertaker as provided under article 4(4) of the GPDO 1995. (See also discussions of both new procedures by Mynors, C. (1995) M'Learned Friend, *Context*, No. 46, June, at pp. 32–3.)

pp. 86–90 Appeal procedures

Further details on inquiry procedures may be found in DoE Circular 24/92. In 1995 the Secretary of State for the Environment announced an intention to review existing inquiry rules which presently may be found in the Town and Country Planning (Inquiries Procedure) Rules 1988 and 1992. An examination of possible issues for reform was made in article published in the Journal of Planning and Environmental Law in February 1996 (Graves, G., Max, R. and Kitson, T. Inquiry Procedure Rules – Another Dose Of Reform [1996] JPL 99).

New legislation has been provided concerning the cost of public inquiries via the Town and Country Planning (Cost of Inquiries etc.) Act 1995.

Chapter 4

pp. 97, 126 Heritage Lottery Fund

By September 1995 nearly £50 million had been awarded in grant funds to 79 projects from the Heritage Lottery Fund. A few large grants have been made including £2.65 million (Highcliffe Castle) and £1.4 million (Tameside Library). In February 1996 the project undertaken by Tyne and Wear Preservation Trust Ltd involving Alderman Fenwick's House in Newcastle upon Tyne (see p. 126) was awarded a grant of £738,678 (*The Journal*, February 14, 1995 p. 13). A high proportion of buildings supported so far by this fund have been churches. However, there remains concern that this source of funding will be used to justify planned cuts in English Heritage's budget (Venning, P. (1995) Lottery Fund Grants So Far, in SPAB News, Vol. 16, No. 4, at p. 1).

p. 97 English Heritage Funding

Indications at the end of 1995 suggests that English Heritage will have reduced budgets for grant aid. Funding assistance may be concentrated more into Conservation Area Partnerships in the future (Sharman, J. (1995) The challenge of tighter budgets, *Conservation Bulletin*, Issue 27, November, at pp. 1–2).

pp. 98–9 Value Added Tax

(See page reference to Chapter 6 below.)

p. 99 Investment performance of listed buildings

The Investment Property Databank published its third review on the investment performance of listed (office) buildings in January 1996. This provided evidence that listed office buildings out-performed all other categories of office buildings in investment terms by almost 2 per cent during 1994. After three years' research, listed office buildings have consistently produced a significantly better performance against other office property. This provides further weight to the argument that listing and commercial viability are not necessarily incompatible. (See Listed Buildings Research Is Published, *Chartered Surveyor Monthly*, February 1996, at p. 61.)

pp. 134–6 Compulsory purchase

A minimum compensation direction was granted in a CPO decision relating to the listed nos. 26–30 Normandy Street, Alton. The direction was significant as the Secretary of State included land for new development within the order which would financially enable the repair of the listed building. (See Steagles, G. (1995) A Rather Useful CPO With Minimum Compensation At Alton, Hampshire, *Context*, No.46, June, at pp. 28–30.)

Chapter 5

p. 141 British Standard Institution draft *Guide to the Care of Historic Buildings*

No further progress has been made on the formulation of a British Standard for building conservation and the draft remains as a guide due to a failure to agree on its contents by members of the committee responsible for drafting and steering its early stages. Concern has been raised that the setting of standards would result in rigid and prescriptive rules. (See Redman, A. (1995)

British Standard Institution Guide to the Principles of Building Conservation: Further developments on this controversial guide, *Building Conservation Journal*, No.13, Winter, RICS Building Conservation Group, at p. 7.)

Chapter 6

Terms of Engagement for Historic Buildings

In 1995 the RICS Building Conservation Group published interim guidance for all Chartered Surveyors providing Building Surveying Services for Historic buildings while existing Conditions of Engagement are under review. The Building Conservation Note 5 *Terms of Engagement For Historic Buildings* provides advice on five stages of engagement:

1. Stage A: Initial Appraisal
2. Stage B: Inspection and Report
3. Stage C: Pre-Contract Services
4. Stage D: Post Contract Services
5. Additional Services

(See *Supplement to Building Conservation Journal*, No.13, Winter 1995, RICS Building Conservation Group.)

pp. 200–202 Value Added Tax

New legislation came into force on 1 January 1996 which effectively closes a loophole in relation to VAT relief on listed buildings. VAT relief is no longer available for alterations made as a consequence of repairs to a listed building (where previously it was possible to apportion, pro rata, the preliminary and access costs between structural alterations and repair work undertaken in a single contract). (See *Building Conservation Journal*, No.13, Winter, RICS Building Conservation Group, at p. 19 and subsequent editorial amendment).

Chapter 7

pp. 217–24 Designation of conservation areas

In *R v. Surrey County Council, ex parte Oakimber Ltd* [1995] EGCS 120 the issue LPA discretion in relation to the designation of conservation areas was raised. Following the decision in *R v. Swansea C.C., ex parte Elitestone Ltd* [1992] JPL 1143 (see p. 222) it was confirmed that LPA's have a broad discretion in designating conservation areas, and in this instance Surrey County Council had not acted irrationally in deciding to designate an area of over 370 acres of land and buildings at Brooklands which it wished to maintain completely intact.

Chapter 8

p. 263–4 Compensation for refusal of Scheduled Monument Consent

Under the fourth requirement which must be satisfied before a claim compensation may be awarded for specified works, where the works were identified as 'not constituting development' or 'authorised by the GDO 1988', the latter should now be read as being authorised by the GPDO 1995.

Chapter 9

pp. 303–6 Register of Historic Parks and Gardens

New arrangements have been made requiring local planning authorities to consult English Heritage on planning applications affecting grade I and grade II* registered sites (via Article 10(1)(o) of the GDPO 1995) and to consult the Garden History Society on applications affecting any registered site, irrespective of their grade (via the Town and Country Planning (Consultation with the Garden History Society) Direction 1995) (see also DoE Circular 9/95). English Heritage and the Garden History Society are now able to advise specifically on proposals affecting designed landscapes, whether or not a particular application also affects the setting of a listed building). (Streeten, A. (1995) Planning consultants for historic parks and gardens, *Conservation Bulletin*, Issue 27, November at p. 20.)

A review of planning issues concerning Historic Parks and Gardens has been published in an article in the Journal of Planning and Environment Law (Lambert, D. Shacklock, V. Historic Parks and Garden: A Review of Legislation, Policy Guidance and Significant Court and Appeal Decisions [1995] JPL 563).

pp. 306–7 Historic battlefields

In June 1995 details of the sites included in the Register of Historic Battlefields were published by English Heritage.

Further information

English Heritage – New development in historic areas

In June 1995 English Heritage published a 'guide to policy, procedure and good practice' entitled *Development in historic areas.* The purpose of this guide is to alert developers and their professional advisers to the advantages 'of trying to reduce avoidable conflict, and the delay and expense it inevitable causes'. The guide is directed at any developer and/or an owner occupier of a listed building who proposes to undertake significant development in relation to a listed building with the first intention of maximising return on investment. This guide also sets out the legislative and policy framework for conservation of the built environment in its widest sense and indicates the responsibilities and roles of both local planning authorities and developers. It provides further information on progressing and carrying out a scheme, and appraising a completed project.

English Heritage – Developing guidelines for the management of listed buildings

This further publication, also released in June 1995, provides a useful discussion on the possibilities for developing management guidelines for listed buildings as a means to encouraging a more rational approach to managing protected buildings. This may be relevant in the context of future user requirements and regarding owner or investor needs for alterations to enable the long-term economic security of listed buildings. The advantages of guidelines is suggested as being to be able to clarify the need for listed building consent, particularly the extent to which minor works can be undertaken without consent. Information is provided on the scope and terms of management guidelines. These include:

- the legal status of the guidelines and the parties involved
- arrangements for review
- the nature of the special interest
- identification of the curtilage and ancillary structures
- the provision and updating of a building archive through recording features and works carried out
- agreed works or areas which are agreed not to affect the character of a building
- works likely to be granted consent
- monitoring of an agreement

The idea of providing such guidelines will necessarily involve consultation of relevant conservation bodies such as the National Amenity Societies and English Heritage as well as public views. By this approach, encouraging a broad agreement of views, the possibility of a legal challenge to any aspect of agreed guidelines between developers/owners and LPAs which indicates that listed building consent is not required for particular works may be reduced.

In summary, English Heritage has encouraged the preparation of such guidelines in appropriate case (i.e. for 'large, relatively modern, statutorily listed commercial or industrial buildings and housing developments ... and other groups of buildings'). However, it is further indicated that LPAs should be aware of the constraints of present law and 'only act within their powers', i.e. they must not fetter their statutory discretion in relation to decisions on the need for listed building consent.

In the future, further scope may be provided for the use of management guidelines by amending current legislation. The debate concerning the need for greater flexibility in listed building legislation is moving towards this end, but time will tell whether new enabling provisions are enacted.

LIST OF ABBREVIATIONS

ACO	Association of Conservation Officers
AHF	Architectural Heritage Fund
AOAI	Areas of Archaeological Importance
APT	Association of Preservation Trusts
BAADLG	British Archaeologists' and Developers' Liaison Group
BPT	Building Preservation Trust
CAC	Conservation Area Consent
CAD	Computer Aided Design
CBA	Council of British Archaeology
CGT	County Gardens Trust
CIP	Calderdale Inheritance Project
COSIRA	Council of Small Industries in Rural Areas
COTAC	Conference on Training in Architectural Conservation
CPO	Compulsory Purchase Order
DHBT	Derbyshire Historic Buildings Trust
DNH	Department of National Heritage
DoE	Department of the Environment
EU	European Union
GDO	General Development Order
GDPO	Town and Country Planning (General Development Procedure) Order 1995 (SI 1995/419)
GHS	Garden History Society
GPDO	Town and Country Planning (General Permitted Development) Order 1995 (SI 1995/419)
ICOMOS	International Council on Monuments and Sites
LBC	Listed Buildings Consent
LBEN	Listed Buildings Enforcement Notice
LOTS	Living Over The Shop
LPA	Local Planning Authority
MAFF	Ministry of Agriculture, Fisheries and Food
MARS	Monuments at Risk Survey
MHLG	Ministry of Housing and Local Government
MPP	Monuments Protection Programme

P(LBCA) 1990	Planning (Listed Buildings and Conservation Areas) Act 1990
PPG	Planning Policy Guidance
RCHME	Royal Commission on the Historical Monuments of England
RIBA	Royal Institute of British Architects
RICS	Royal Institution of Chartered Surveyors
SMC	Scheduled Monument Consent
SMR	Sites and Monuments Record
SPAB	Society for the Protection of Ancient Buildings
TPO	Tree Preservation Order
VAT	Value Added Tax
WHS	World Heritage Site

the 1979 Act	Ancient Monuments and Archaeolgical Areas Act 1979
the 1990 Regulations	Town and Country Planning (Listed Buildings and Buildings on Conservation Areas) Regulations 1990
the principal planning Act	Town and Country Planning Act 1990

LEGAL INDEX

Table of cases

Table of statutes

Table of statutory instruments

INDEX

UNIVERSITY OF WOLVERHAMPTON
LEARNING RESOURCES